Machine Design for Technology Students

A Systems Engineering Approach

Synthesis Lectures on Mechanical Engineering

Synthesis Lectures on Mechanical Engineering series publishes 60–150 page publications pertaining to this diverse discipline of mechanical engineering. The series presents Lectures written for an audience of researchers, industry engineers, undergraduate and graduate students.

Additional Synthesis series will be developed covering key areas within mechanical engineering.

Towards Analytical Chaotic Evolutions in Brusselators
Albert C.J. Luo and Siyu Guo
2020

Modeling and Simulation of Nanofluid Flow Problems
Snehashi Chakraverty and Uddhaba Biswal
2020

Modeling and Simulation of Mechatronic Systems using Simscape
Shuvra Das
2020

Automatic Flight Control Systems
Mohammad Sadraey
2020

Bifurcation Dynamics of a Damped Parametric Pendulum
Yu Guo and Albert C.J. Luo
2019

Reliability-Based Mechanical Design, Volume 2: Component under Cyclic Load and Dimension Design with Required Reliability
Xiaobin Le
2019

Reliability-Based Mechanical Design, Volume 1: Component under Static Load
Xiaobin Le
2019

Solving Practical Engineering Mechanics Problems: Advanced Kinetics
Sayavur I. Bakhtiyarov
2019

Natural Corrosion Inhibitors
Shima Ghanavati Nasab, Mehdi Javaheran Yazd, Abolfazl Semnani, Homa Kahkesh, Navid Rabiee, Mohammad Rabiee, and Mojtaba Bagherzadeh
2019

Fractional Calculus with its Applications in Engineering and Technology
Yi Yang and Haiyan Henry Zhang
2019

Essential Engineering Thermodynamics: A Student's Guide
Yumin Zhang
2018

MEMS Barometers Toward Vertical Position Detection: Background Theory, System Prototyping, and Measurement Analysis
Dimosthenis E. Bolanakis
2017

Engineering Finite Element Analysis
Ramana M. Pidaparti
2017

Machine Design for Technology Students: A Systems Engineering Approach
Anthony D'Angelo, Jr.

ISBN: 978-3-031-79684-5   paperback
ISBN: 978-3-031-79685-2   ebook
ISBN: 978-3-031-79687-6   hardcover

DOI 10.1007/978-3-031-79685-2

A Publication in the Springer series
SYNTHESIS LECTURES ON MECHANICAL ENGINEERING

Lecture #33
Series ISSN
Print 2573-3168   Electronic 2573-3176

Machine Design for Technology Students

A Systems Engineering Approach

Anthony D'Angelo, Jr.

SYNTHESIS LECTURES ON MECHANICAL ENGINEERING #33

ABSTRACT

This book is intended for students taking a Machine Design course leading to a Mechanical Engineering Technology degree. It can be adapted to a Machine Design course for Mechanical Engineering students or used as a reference for adopting systems engineering into a design course. The book introduces the fundamentals of systems engineering, the concept of synthesis, and the basics of trade-off studies. It covers the use of a functional flow block diagram to transform design requirements into the design space to identify all success modes.

The book discusses fundamental stress analysis for structures under axial, torsional, or bending loads. In addition, the book discusses the development of analyzing shafts under combined loads by using Mohr's circle and failure mode criterion. Chapter 3 provides an overview of fatigue and the process to develop the shaft-sizing equations under dynamic loading conditions.

Chapter 4 discusses power equations and the nomenclature and stress analysis for spur and straight bevel gears and equations for analyzing gear trains. Other machine component topics include derivation of the disc clutch and its relationship to compression springs, derivation of the flat belt equations, roller and ball bearing life equations, roller chains, and keyways.

Chapter 5 introduces the area of computational machine design and provides codes for developing simple and powerful computational methods to solve: cross product required to calculate the torques and bending moments on shafts, 1D stress analysis, reaction loads on support bearings, Mohr's circle, shaft sizing under dynamic loading, and cone clutch.

The final chapter shows how to integrate Systems Engineering into machine design for a capstone project as a project-based collaborative design methodology. The chapter shows how each design requirement is transformed through the design space to identify the proper engineering equations.

KEYWORDS

systems engineering, machine design, mechanical engineering technology, functional analysis, functional flow block diagrams, requirement analysis, computational machine design, fatigue analysis, capstone project

Contents

Preface

The idea for this book was not from an epiphany but from my notes used in teaching a Machine Design for Technology course leading to an Applied Associates degree in Mechanical Engineering Technology. The focus of the book is two-fold.

1. "Machine:" Covers the fundamentals of stress analysis when structures experience axial, torsional, or bending loads. In addition, when a combination of loading conditions exists on a structure, students learn to use Mohr's Circle and the von Mises failure criteria. Chapter 2 reviews the basics of stress analysis. Chapter 3 covers fatigue and the painstaking algebra to derive the shaft diameter equation. I opted to include this as a separate chapter for two reasons. One, to show students that equations do not magically appear in textbooks but are derived. The second reason, and more important, is to impart to students that extreme care in using design equations and understanding their assumptions and limitations. Chapter 4 covers the basic stress and design analysis equations for standard machine components.

2. "Design:" Performing complex calculations by hand or using computer-based software is only one part of a Machine Design course. Chapter 1 covers the basics of Systems Engineering (SE). The chapter introduces the eight-phase SE model that creates the foundation of developing a disciplined approach in designing complex systems. Students learn how transform system requirements, through the use of functional analysis, to design requirements. Chapter 6 presents a case study example for students, given only eight requirements, to use systems engineering and machine component analysis to design a capstone conveyor system. Finally, during the process of creating examples and solving equations I decided to add Chapter 5: Computational Machine Design. The chapter shows the process and code to simplify and solve various topics.

Students using this book should have a working understand of drawing Free Body Diagrams. In addition, students should have taken a course in stress analysis or mechanics of materials. Many references are available that develop the underlying theory and assumptions for stress equations. It is incumbent on the student to understand the underlying assumptions. The book uses the limit state function concept to solve for stresses: $\sigma_{design} < \sigma_{allowable}$

Fundamentals are a key focus in this book and analysis of machine components is limited to one or two component types. For example, analysis of gears is limited to spur and straight bevels. Bearings are limited to the ball and roller type. Shaft equations use the Soderberg criterion. By limiting the number of gears to spur and straight bevel types and bearings to ball or roller, students are able to focus on the design aspect as well as the analysis of machine components. Finally, SE is a disciplined approach to designing complex systems. At an undergraduate level, SE is in its infancy. Nevertheless, Chapter 6 walks through the design of a capstone project using only eight requirements. The design requirements include manufacturing, operation, and maintenance. The design requirements, in some cases, are intentionally vague so students learn to challenge the requirements and apply Chapter 1 concepts in brainstorming, synthesis, and trade studies, that are key functions for the "art" of design.

Anthony D'Angelo, Jr.
October 2020

Acknowledgments

It is rare to get an opportunity to develop your set of notes into a manuscript. It is easy for someone to assume you just take a set of notes and transform them. However, I could not accomplish the manuscript without the support and encouragement from those who have or had a strong influence in my life. I start with my parents who laid the foundation early and taught me to work hard, stay focused, follow your passion, and failure is not weakness but a process to learn. Their support and encouragement I will never forget. I need to acknowledge three very important people. First, to my son Alex whose creativity and view of the world is through his lens. He has taught me to focus on the details, all while never forgetting the landscape. His attention to detail to tell a story through his photographs inspires me to write through figures and drawings and not to keep my focus on the equations. Next, to my daughter Keri and her dogged Ph.D. pursuit in math to understand and uncover new math concepts in its truest form has taught me to present the material in its most fundamental manner. She has made me understand that the most complex topics need to be broken down to their foundation. Last, but not least, to my wife and best friend Nina whose unwavering love, support, and encouragement I can never measure. Her creativity and gift of making any topic easily understood has been my inspiration and sprinkled throughout this manuscript. I am forever grateful for their support and love.

Anthony D'Angelo, Jr.
October 2020

CHAPTER 1

Systems Engineering

1.1 INTRODUCTION

Systems engineering (SE) brings together a mutli-disciplined team to develop a new (or modify an existing) and complex system using a systematic modeling approach. Many systems engineering models exists, but we employ the model shown in Figure 1.1. SE starts with a need for a new system. Next, the SE process transitions to the Exploration phase. During this phase, teams use brainstorming techniques to generate as many systems or ideas as possible to meet the customers' requirements. The Definition phase includes performing a trade study to recommend the final concept. The Design phase is where engineering occurs to perform stress analysis and machine component analyses. Once the design is completed, the next stage is test and evaluation. Finally, the last stage of the SE model is Post Development and concentrates on how the system is built and its operation and support.

1.2 SE MODEL

Needs Analysis Phase: A new requirement or implementing new technology into an existing design is the driving force for conducting a needs analysis. In the SE model, this is the first step in developing a design. However, in practice, designers must convince the decision makers that development of the proposed system has a reasonable cost, an acceptable risk level, and at least one high-level conceptual design exists.

Concept Exploration Phase: During this phase students choose their design teams and begin to develop ideas that meet the design requirements. Brainstorming is one technique where a group gets together, in an informal manner, to generate ideas and/or solutions to a problem. Members must follow certain rules, such as no idea will be discarded, no idea from any member must be criticized or ridiculed, and all ideas must be recorded.

Example 1.1 (Brainstorming)
Students are broken into teams and asked to develop a minimum of five solutions to the following problem: their instructor must find a way to get from his home, to work, and to class. The distance from home to work is 40 miles, from work to college is 10 miles, and the college is between the workplace and instructor's home (Figure 1.2). Table 1.1 records various concepts.

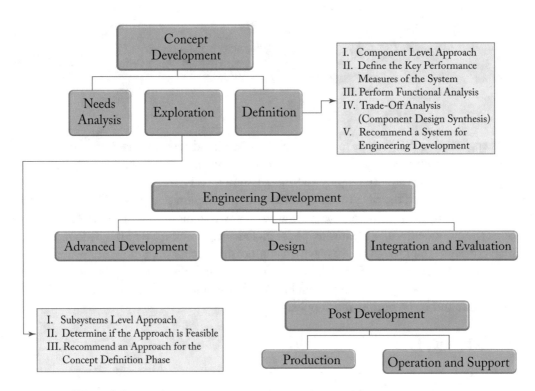

Figure 1.1: SE model.

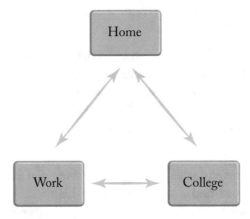

Figure 1.2: Example 1.1 brainstorming.

Table 1.1: Ideas

Team Member	Ideas	
1	Drive car/truck	Drone technology
2	Walk	Hitch a ride
3	Motorcycle	Ride share program
4	Bus	
5	Train	

Figure 1.3: Example 1.2 island hopping.

Table 1.2: Ideas

Team	Ideas	
Member 1	Build a bridge	Drain the lake and walk
Member 2	Ferry	
Member 3	Helicopter	
Member 4	Freeze the lake and walk	

Example 1.2   (Island)
Students are broken into their teams and asked to develop a minimum of five solutions to the following problem: how do I get across the lake from Island 1 to 2 (Figure 1.3)? Table 1.2 records various concepts.

Concept Definition Phase: This phase analyzes all of the feasible concepts discussed during the concept exploration phase and chooses one of the concepts. Using a trade study process is necessary to choose a design to recommend to the decision makers. Figure 1.4 demonstrates how a trade study is conducted in practice. This example uses reliability as the performance function. For example, for Performance 7 under the Criteria column design alternative 1 has a reliability value = 0.6 compared with a reliability = 0.9 for design alternative 4. But, design alternative 1 is better on cost and weight compared to design 4. We must make a non-bias decision for choosing an alternative. A trade study tool is demonstrated, to the teams, showing 5 design concepts evaluated against 17 criteria. Figure 1.5 shows final rankings. For our example, we chose design concept 4 (highest value in the last column).

	W_i		Criteria	Design Alternatives					Criterion Type	Parameters
	Raw	Normalized		1	2	3	4	5		
0.12766	6	0.127659574	Weight	24000	23000	26000	27500	22000	VI	$\sigma = 1000$
0.106383	5	0.106382979	Cost	9	7.5	10.5	12	6	IV	$q = .1, p = .3$
0.085106	4	0.085106383	Performance 1	0.84	0.92	0.76	0.88	0.85	III	$p = .2$
0.042553	2	0.042553191	Performance 2	0.75	0.7	0.75	0.95	0.8	VI	$\sigma = 0.5$
0.06383	3	0.063829787	Performance 3	0.72	0.9	0.8	0.95	0.9	II	$q = .2$
0.021277	1	0.021276596	Performance 4	0.85	0.85	0.9	0.92	0.9	V	$q = .1, p = .2$
0.06383	3	0.063829787	Performance 5	0.8	0.87	0.8	0.8	0.8	VI	$\sigma = 0.5$
0.085106	4	0.085106383	Performance 6	0.6	0.8	0.77	0.8	0.6	IV	$q = .1, p = .3$
0.042553	2	0.042553191	Performance 7	0.6	0.87	0.9	0.9	0.6	III	$p = .2$
0.042553	2	0.042553191	Performance 8	0.6	0.55	0.9	0.75	0.6	VI	$\sigma = 0.5$
0.06383	3	0.063829787	Performance 9	0.8	0.87	0.7	0.88	0.8	II	$q = .2$
0.021277	1	0.021276596	Performance 10	0.65	0.87	0.65	0.8	0.65	V	$q = .1, p = .2$
0.042553	2	0.042553191	Manufacturing	0.88	0.75	0.8	0.92	0.88	VI	$\sigma = 0.5$
0.06383	3	0.063829787	Test & Evaluation	0.84	0.87	0.9	0.9	0.9	IV	$q = .1, p = .3$
0.06383	3	0.063829787	Operational 1	0.8	0.72	0.6	0.9	0.8	III	$p = .2$
0.021277	1	0.021276596	Operational 2	0.7	0.87	0.9	0.8	0.6	VI	$\sigma = 0.5$
0.042553	2	0.042553191	Operational 3	0.6	0.6	0.45	0.9	0.6	VI	$\sigma = 0.5$

Figure 1.4: Trade study matrix.

	a_1	a_2	a_3	a_4	a_5	ϕ^+	ϕ^-	$\phi^+ - \phi^-$
a_1	0	0.18398375	0.06624344	0	0.21718701	0.46741421	0.731832	-0.26442
a_2	0.24437059	0	0.19823961	0.10721295	0.1899094	0.73973254	0.588963	0.15077
a_3	0.20413471	0.14871671	0	0.10867673	0.2232378	0.68476596	0.506644	0.178122
a_4	0.17694329	0.14966703	0.13577755	0	0.17722251	0.63961038	0.322273	0.317338
a_5	0.10638298	0.10659521	0.10638298	0.10638298	0	0.42574415	0.807557	-0.38181
ϕ^-	0.73183157	0.58896271	0.50664358	0.32227266	0.80755672			

Figure 1.5: Final rankings.

Technology Development Phase: This phase develops critical technology. For example, if a new battery for autonomous vehicles is required or new materials to meet a lightweight and high-strength criterion.

Design Phase: The heart of engineering and machine design analyses takes place in this phase. Students have chosen the design and begin to develop detailed design drawings. Hand calculation as well as computer modeling and simulation are critical during this phase. Students use stress analysis, fatigue, and machine component analysis to ensure the final design meets all user and engineering code requirements. Emphasis at this junction should be to look back and forward. Students learn, during this phase, to challenge requirements as well as understand how the design will be built and how the customer will operate and maintain the system.

Integration and Evaluation Phase: Does the design work? Does the design meet the customer's requirements? During this phase, prototypes of the design are built and tested to validate the design. Simulation methods, such as finite element analysis or discrete event simulations, are acceptable alternatives in lieu of physical testing.

Production Phase: At this point, the production facility is ready to build the final design. Students are taught that, when possible, modular designs, common manufacturing tools and equipment, and off-the-shelf parts and materials should be used. Design changes at this phase are expensive and will result in cost overruns and schedule delays.

Operation and Maintenance Phase: The customer has taken delivery of the final design. The customer has responsibility for maintenance and operation of the design but it is imperative that students understand the operation and maintenance requirements during the design phase.

1.3   FUNCTIONAL ANALYSIS: WHAT vs. HOW?

The key to designing a complex system starts with the question of what is the system supposed to do? A function is defined as a "verb-object." For example, study for exam or turn-in assignment. For the examples, the verbs are *study* and *turn-in*. Students should study for an exam but "how" they study is a personal choice. The same is true for turning in their assignment. The verb is "turn-in" but the student may wish to attend class and personally hand in a written copy, use Blackboard, or simply e-mail.

During the Concept Definition stage of the SE process we want to define what the system must do and not how it accomplishes the task. Figure 1.6 shows a Functional Flow Block Diagram (FFBD) for designing a transmission system. The FFBD is a sequential flow of tasks starting at the 1st level. In this example, we identify the need for a transmission, brainstorm ideas (gears, belts, chains, etc.), choose the design, develop any technology required, design, test, build, and maintain. We create as many levels as required until we identify all functions. The "Design Transmission" function is broken to the next level to design the power system, the transfer system, and finally the output system. Level 3 illustrates the power system function by choosing the motor, deigning a clutch, and designing the input shaft component.

The FFBD is a disciplined and sequential method that outlines the function of the system. However, it does not link the customer's requirements to a specific function. The Integration Definition Function Modeling (IDEF0) [1] technique ties the requirement or Technical Performance measures (TPMs) to a specific function. Figure 1.7 illustrates the IDF0 model its four components.

1. Control: The TPMs. For example, when choosing a bearing based on life the operating conditions of the conveyor are given.

2. Mechanisms: Students are the analysts and choose the proper equations and software to analyze machine components.

3. Inputs: The choice of parts under consideration. For example, students can choose between a cone or disc clutch.

Functional Flow Block Diagram

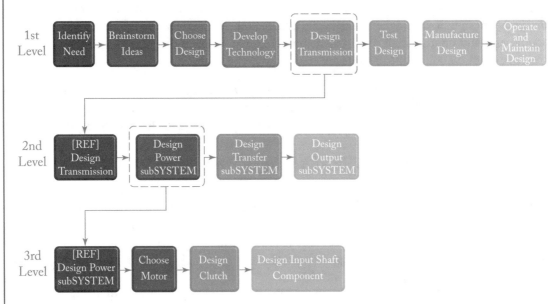

Figure 1.6: **FBD** for transmission design.

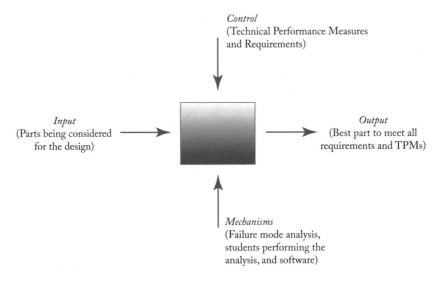

Figure 1.7: **IDEF0** model.

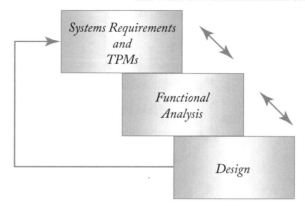

Figure 1.8: FUSE-FFBD model.

4. Output: The final choice that bests meets the TPMs, systems requirements, and cost.

The Function-based Systems Engineering (FUSE) model [2], shown in Figure 1.8, links the TPMs and requirements to the IDEF0 model, and design analyses.

The FUSE-FFBD model has a unique feature and one students must keep in mind. The model is reversible. Students should challenge the TPMs and requirements if a requirement is overly restrictive or a cheaper alternative is available that does not meet a specific TPM but does not affect the overall performance.

Example 1.3   (IDEF0 Modeling)
A customer has a requirement for a decorative bracket to hold hanging plants at a height of 10 feet weighing up to 10 pounds. Set up the IDEF0 and FUSE Model. TPMs: Must hold max 10 pounds. Decorative (Customer Requirement).

Using the IDEF0 Model:

Control: Max vertical weight to hold is 10 pounds. (TPM given by the customer.)

Mechanisms: Analyst. Static equations. Strength of material equations. Calculator. (Required equations and calculator to solve equations as well as who is performing the analysis.)

Inputs: The bracket designs under consideration.

Output: The final bracket that meets customer's requirements.

Using the FUSE Model (Figure 1.9):

The TPM is the customer requirement to hold 10 pounds. Students should note the requirement "Decorative" is ambiguous and the designer should challenge it or ask for clarification.

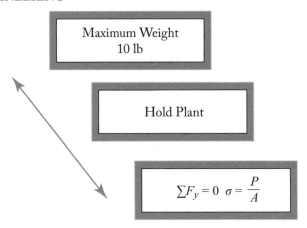

Figure 1.9: FUSE model for bracket.

1.4 CONCEPT EXPLORATION TO CONCEPT DEFINITION TO DESIGN (THE PROCESS)

To get from the chosen design to the final product, we introduce the systems engineering hierarchy. As shown in Figure 1.10, the System is the top level followed by major subSystems, components, subComponents, and parts. The design process takes place at the parts level. For example, a spur gear, shaft, and keyway make a subcomponent. However, during the design phase the keyway, shaft, and gear are parts and must be analyzed separately using the appropriate equations and assumptions.

In the design phase all parts are analyzed. To ensure the success of the design, the design space identifies all of the success modes that ensures the final design.

1. Meets and accounts for all of the requirements and TPMs.

2. Identifies all the parts making up the system and assigns the proper machine design engineering equations.

3. Identifies any missing requirements.

4. Does not add parts or features not required by the customer. This is sometimes referred to as "gold platting."

An FFBD is required to develop the design space and identify all success modes. Figure 1.11 constructs the design space starting with the FFBD. The 1st level identified that a transmission must be developed. The transmission represents the system. The 2nd level identified three major subSystems:

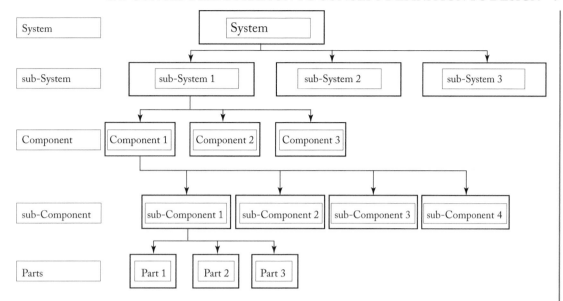

Figure 1.10: Design hierarchy....from system to parts.

1. Power

2. Transfer

3. Output.

The 3rd level represents the design process for the Power subSystem. Using a clutch is a requirement for the system. The student should note that the function is "Design Clutch" and the type of clutch is unknown. The main function of a clutch is to "Transmit Torque." Given that a clutch is a requirement, students need to identify what makes a clutch work. This book uses the term "success mode" or "success event" to identify the proper engineering methods and equations. This SE approach allows the final design meting all the customer requirements. The customer wants to know if a delivered system will perform to the given requirements and not why it failed.

Example 1.4 (Clutch design)
Function: (verb-object) Design Clutch. This just describes what the designer is supposed to accomplish.

Success Event: Transmit Torque. What is the clutch supposed to do is described at this point.

Functional Flow Block Diagram

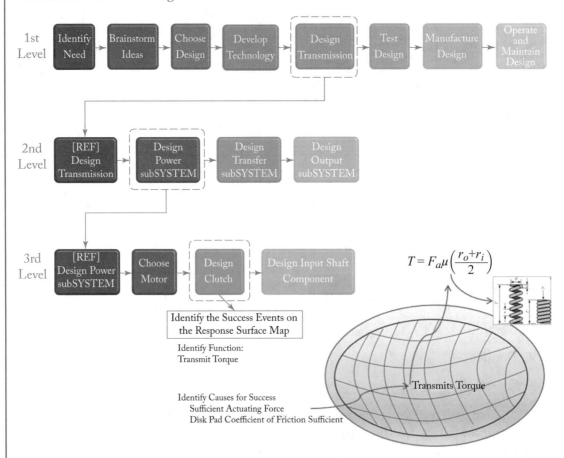

Figure 1.11: Clutch design.

Causes for Success:

1. Sufficient actuating force.

2. Disc pad coefficient of friction is adequate.

   Identified are the proper torque equation (disc clutch) and proper compression spring equations.

Using this method allows the designer to account for all customer requirements. If a requirement is missing or ambiguous the designer can get clarification. The FUSE model with the IDEF0 identifies the specifications, requirements, proper equations, the analysis, and all the required software to properly design the system.

CHAPTER 2

Basics of Mechanics of Materials

2.1 INTRODUCTION

This chapter presents a review of statics and strength of materials and students should have completed courses in both topics prior to undertaking machine design. This chapter does not discuss all of the topics taught in a traditional stress analysis course and limits stress calculations to shafts with circular cross sections. For students to be successful in machine design they must master the concept of a Free Body Diagram (FBD). Students must understand the type of load applied to a structure as well as the load's magnitude and direction. The approach taken begins with a single load causing an axial stress. Next, we show how torque causes shear stresses in a structure. Finally, we introduce bending stresses in shafts. Once the student understands the individual pieces, we put it all together through the use of Mohr's circle and failure theories. The focus is on von Mises failure theory and Chapter 3 develops the shaft diameter equation using the topics discussed in this chapter.

2.2 THE FREE BODY DIAGRAM

Students must create accurate FBDs to be successful in a machine design course. It is critical for students to understand the type of loading and its orientation. This section covers the FBD from an SE perspective.

Example 2.1 (FBD)
A customer requires a bracket capable of holding a 2500-lb axial load. Figure 2.1 illustrates the chosen design using three parts (bar, pin, and bracket) that meets the customer's requirements. Figure 2.2 is an exploded view of the design and Figure 2.3 is the FBD for each part.

Once the FBD is completed students can incorporate the FUSE-FFBD model (Figure 2.4) to create the design space. The model, discussed in Chapter 1, identifies the requirement (2500-lb load), the function (hold load), and the equations, software, and analysts required.

Example 2.2 (FUSE-FFBD)
Create the design space for the FBD in Example 2.1.

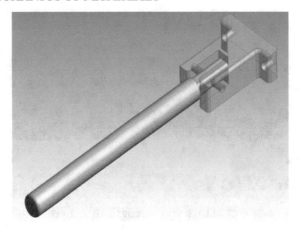

Figure 2.1: Bracket.

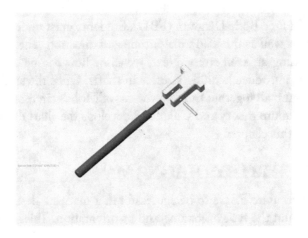

Figure 2.2: Bracket exploded view.

Design Space (the success modes identified from FBD):

1. Bar holds axial load.

2. Shear pin works in double shear.

3. Bar capable of withstanding bearing stress.

4. Bracket capable of withstanding bearing stress at hole.

5. Bar does not have excessive deformation.

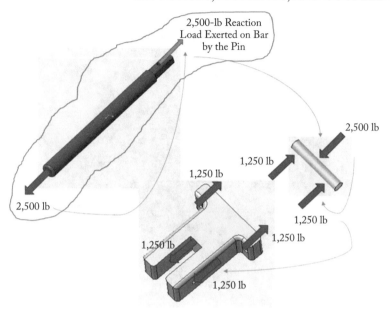

Figure 2.3: Bracket FBD.

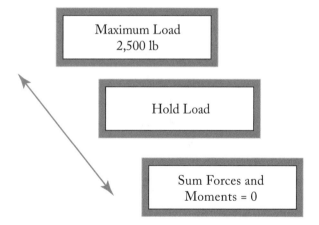

Figure 2.4: FUSE model for bracket.

2.3 FORCES, MOMENTS, AND COUPLED SYSTEMS

Equation (2.1) are the equilibrium equations. A coupled system is illustrated in Figure 2.5 and described mathematically using Equations (2.2), (2.3), and (2.4). Students should review and master these concepts because they play a critical role in determining the forces and torques

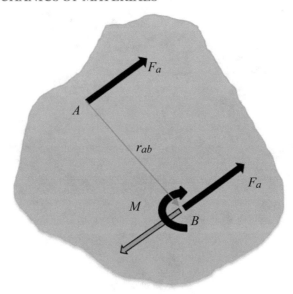

Figure 2.5: Coupled system.

applied to a shaft.

$$\sum F_x = 0$$
$$\sum F_y = 0$$
$$\sum F_z = 0$$
$$\sum M_x = 0$$
$$\sum M_y = 0$$
$$\sum M_z = 0.$$

(2.1)

$$M = Fd$$

(2.2)

$$\vec{M} = \vec{r} X \vec{F} = \begin{bmatrix} i & j & k \\ r_x & r_y & r_z \\ F_x & F_y & F_z \end{bmatrix}$$

(2.3)

$$\vec{M} = (r_y F_z - r_z F_y)\hat{i} - (r_x F_z - r_z F_x)\hat{j} + (r_x F_y - r_z F_x)\hat{k}.$$

(2.4)

Example 2.3   (Cross Product)
Figure 2.6 shows a simple shaft with one spur gear attached. The shaft is supported by two ball bearings. Find the loads and moments acting on the shaft and the reaction load and on the left ball bearing support.

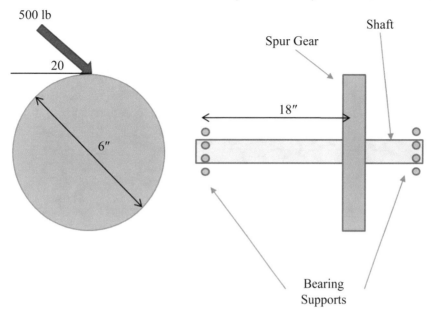

Figure 2.6: Spur gear-shaft analysis.

$$\vec{r} = 18\hat{i} + 3\hat{j}$$
$$\vec{F} = -500\sin 20\hat{j} + 500\cos 20\hat{k}.$$

Cross-product torque and bending moment analysis computational machine design computer code

Moment is returned representing:

```
Torque=Mx;
Bending yAXIS=My;
Bending zAXIS=Mz;
function [Moment] = DistForce( rx,ry,rz,Fx,Fy,Fz)
Mx= (ry*Fz)-(rz*Fy);
My= -((rx*Fz)-(rz*Fx));
Mz= (rx*Fy)-(ry*Fx);
Moment=[Mx,My,Mz];
end
```

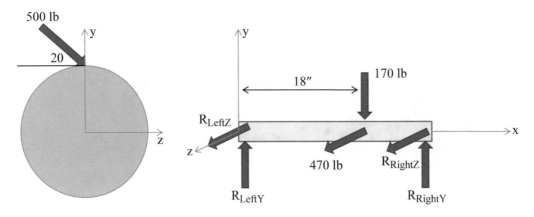

Figure 2.7: FBD for spur gear problem (3D).

RESULTS:

Moments $=$ DistForce(18,3,0,0,$-$170,470)

Results $\vec{M} = \vec{r} \times \vec{F}$

Torque $= 1410$ in-lb

Bending yAXIS $= -8460$ in-lb

Bending zAXIS $= -3060$ in-lb

Another approch is to develop the FBD given in Figure 2.7. Keep in mind that a spur gear attached to a shaft is a 3D problem.

Moment about z-axis is $M_z = -(170)(18) = -3060$ in-lb.

Moment about y-axis is $M_y = -(497)(18) = -8460$ in-lb.

The Torque, $T = (470)(3) = 1410$ in-lb.

Both methods give the same results. Although it is simpler to use the cross-product and computer model, the FBD illustrates the loads and moments imparted on the shaft by the spur gear. Later on you will see how this plays a crucial role in sizing the shaft.

Example 2.4 (Couple)

Use the couple technique to show the force and torque imparted by the gear on the shaft.

Figure 2.8 illustrates the sequence.

Step 1: Add equal and opposite forces at the shaft's center.

Step 2: Replaces the applied 500-lb load and equal and opposite force at the center with the moment ($500 \cos 20 = 1410$ in-lb).

Example 2.5 (Vector Cross Product)

Given point **a** at located at coordinates (3,5,1) and vector $\vec{F}_a$ with a magnitude of 1000 lb and

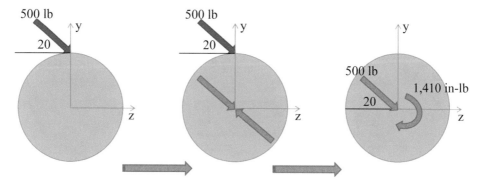

Figure 2.8: Force couple sequence.

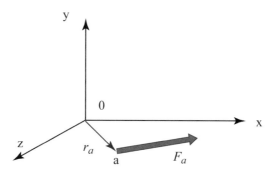

Figure 2.9: Vector cross product.

acting along a line containing the points (3,5,1) and (12, 10, 8), find the Moment about the origin. Equation (2.5) defines the unit vector $\hat{u}$. Using this equation the force vector, $\vec{F}$, is defined using Equation (2.6):

$$\hat{u} = \frac{\vec{r_a}}{\sqrt{r_x^2 + r_y^2 + r_z^2}} = \frac{r_x\hat{i} + r_y\hat{j} + r_z\hat{k}}{\sqrt{r_x^2 + r_y^2 + r_z^2}} \tag{2.5}$$

$$\vec{F} = F\hat{u_F} \tag{2.6}$$

$$\hat{u_F} = \frac{(12-3)\hat{i} + (10-5)\hat{j} + (8-1)\hat{k}}{\sqrt{9^2 + 5^2 + 7^2}}.$$

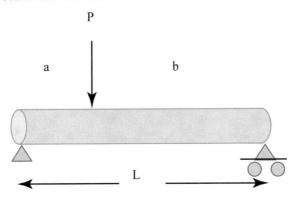

Figure 2.10: Simply supported beam.

$$\vec{F} = 1000\frac{(9)\hat{i} + (5)\hat{j} + (7)\hat{k}}{\sqrt{155}} = 722.9\hat{i} + 401.6\hat{j} + 526.3\hat{k}$$

$$\vec{M} = (r_y F_z - r_z F_y)\hat{i} - (r_x F_z - r_z F_x)\hat{j} + (r_x F_y - r_z F_x)\hat{k}$$

$$\vec{M} = ((5)(562.3) - (1)(401.6))\hat{i} - ((3)(562.3)$$
$$- (1)(722.9))\hat{j} + ((3)(401.6) - (5)(722.9))\hat{k}$$

$$\vec{M} = 2409.9\hat{i} - 964.0\hat{j} - 2409.7\hat{k}.$$

2.4   CIRCULAR CROSS-SECTION BEAMS: A PIECEWISE APPROACH TO SHAFT DESIGN ANALYSIS

This section shows the steps required to calculate the reaction loads in the y- and z-directions acting on beam as well introducing torque diagrams. We discuss the stresses induced on shafts under multiple loads. In Chapter 4 we show where the loads arise, i.e, gears. belts, etc. Why circular cross sections? Because shafts are treated as beams with circular cross-section. We start with Figure 2.10, a simply supported circular cross-section beam of length L subjected to a force P located at a distance a. It is critical that student's master the FBD as shown in Figure 2.11 and apply the equilibrium Equations (2.1) to solve for the reaction loads. Proper calculation of the reaction loads are required for two reasons: (1) shear and bending moment diagrams are required for stress analysis and (2) we will learn a little later that the reaction loads are the radial loads required to properly size bearings.

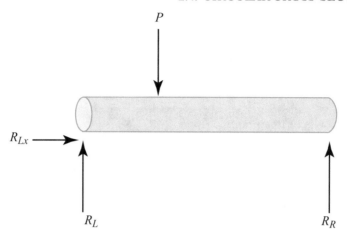

Figure 2.11: **FBD for simply supported beam.**

$$\rightarrow + \sum F_x = 0$$
$$= R_{Lx} = 0$$
$$\uparrow + \sum F_y = 0$$
$$= R_L + R_R - P = 0$$
$$\circlearrowleft + \sum M_z = 0$$
$$= R_R L - Pa = 0$$
$$R_R = \frac{Pa}{L}$$

substituting and solve for R_L yields:

$$R_L = P - R_R = P - \frac{Pa}{L} = \frac{P(L-a)}{L} = \frac{Pb}{L}.$$

Example 2.6   (Calculate the Reaction Loads at beam supports)
Given a 5000-lb load located 8″ from the left end, find the left and right reaction loads.

$$R_R = \frac{Pa}{L} = \frac{(5000)(8)}{20} = 2000.0 \, \text{lb}$$
$$R_L = \frac{Pb}{L} = \frac{(5000)(12)}{20} = 3000.0 \, \text{lb}.$$

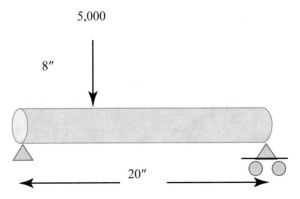

Figure 2.12: Example 2.4 with 5000-lb load.

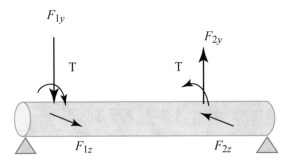

Figure 2.13: Complex shaft loads.

One thing students should observe before continuing with the design or analysis is do the results look OK? Observe the answer and notice that the 5000-lb load is close to the left reaction pin. So? I should expect the left reaction, $R_L > R_R$.

The previous example was a single concentrated load. In practice, shafts are subjected to complex loading under dynamic conditions, as illustrated in Figure 2.13.

It is best to reduce the complex loads into load simple cases. Figure 2.14 shows the FBD for the shaft illustred in Figure 2.13. To calculate the stresses it is best to break the FBD into simple pieces by performing the following steps.

Step 1. Draw the FBD diagram, as illustrated in Figure 2.15 for loads acting only in the y-direction. Using our equations of equilibrium we can solve for the reaction loads in the y-direction.

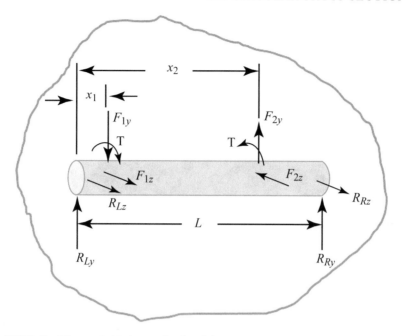

Figure 2.14: **FBD** for Figure 2.13 (complex loads).

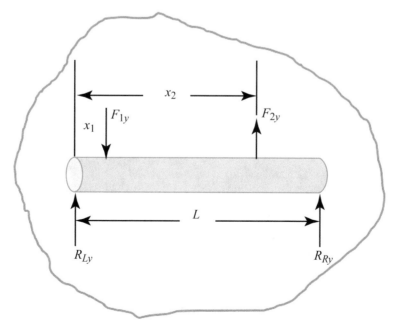

Figure 2.15: **FBD** $x-y$ plane.

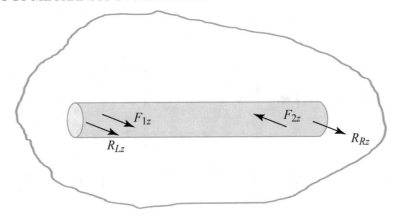

Figure 2.16: **FBD** y–z plane.

$$\uparrow + \sum F_y = 0$$
$$= R_{Ly} + R_{Ry} - F_{1y} + F_{2y} = 0$$
$$\circlearrowleft + \sum M_z = 0$$

$$R_{Ry} = \frac{-F_{1y}x_1 + F_{2y}x_2}{L}$$
$$R_{Ly} = -R_{Ry} + F_{1y} - F_{2y}.$$

Step 2. Draw the FBD diagram, as illustrated in Figure 2.16 for loads acting only in the z-direction. Repeat the process using our equations of equilibrium and solve for the reaction loads in the z-direction.

$$\nearrow + \sum F_z = 0$$
$$= R_{Lz} + R_{Rz} + F_{1z} - F_{2z} = 0$$
$$\circlearrowleft + \sum M_y = 0$$

$$R_{Rz} = \frac{-F_{1z}x_1 + F_{2z}x_2}{L}$$
$$R_{Ly} = -R_{Rz} - F_{1z} + F_{2z}.$$

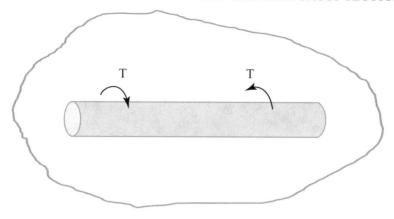

Figure 2.17: FBD constant torque.

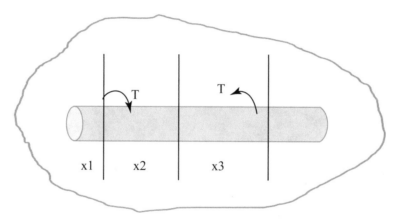

Figure 2.18: Cuts along shaft.

Step 3. Draw the FBD diagram, as illustrated in Figure 2.17 for the torque, T, acting along the beam's axis. Here we treat the torque FBD a little differently than applied loads because students often confuse internal torque to rotation. A shaft can rotate without having an internal torque reaction. To illustrate this concept, slice the beam at three points along its length and calculate the internal torque, T', in each section. Figure 2.18 illustrates where the three slices are taken. The first slice, section x_1, is at any point along the beam before the first torque is applied. What results is the FBD and internal torque, T', illustrated in Figure 2.19. Solving for T' shows there is no internal torque along any point on the beam when $x < x_1$:

$$\circlearrowleft + \sum T = 0$$

$$T' = 0 \, (0 < x \leq x_1).$$

Figure 2.19: Cut at x_1.

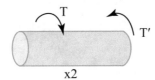

Figure 2.20: Cut at x_2.

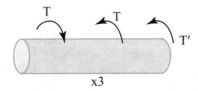

Figure 2.21: Cut at x_3.

The second slice (section x_2) shown in Figure 2.20 occurs along the beam's axis where $x_1 < x < x_2$. Again, solving for T' shows the internal torque along any point on the beam when $x_1 \leq x < x_2$ is just the applied torque, T:

$$\circlearrowleft + \sum T = 0$$
$$T - T' = 0$$

$$T' = T \; (x_1 \leq x < x_2).$$

Finally, Figure 2.21 (section x_3) slices the beam at a point x ($x_2 \leq X < L$) and shows the $T' = 0$. Using this process we construct the Torque Diagram shown in Figure 2.22. It is critical to understand how to construct the torque diagram and in the next chapter we will show its importance to the shaft sizing equation:

$$\circlearrowleft + \sum T = 0$$
$$T - T - T' = 0$$

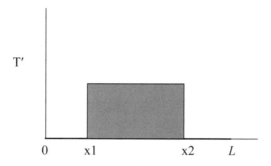

Figure 2.22: Torque diagram.

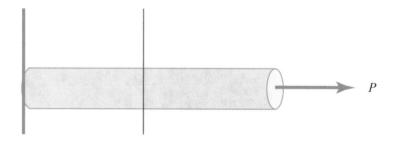

Figure 2.23: Axial loaded shaft.

$$T' = 0 \ (x_2 \geq x < L).$$

2.5   STRESS REVIEW

Section 2.5 covers the fundamentals of stress analysis. Discussion is limited to axial, shear, and bending stresses with emphasis on developing the stress equations through the application of calculus and discussing the assumptions and limitations of each equation.

2.5.1   AXIAL LOAD

Starting with axial stress, Figure 2.23 shows a simple shaft under axial loading. The FBD is illustrated in Figure 2.24 to show how Equation (2.7) $\sigma = \frac{P}{A}$ is derived. Equation (2.7) is the equation for axial stress:

$$\rightarrow + \sum F_x = 0$$
$$= P - \sigma dA = 0$$

Figure 2.24: Internal stress distribution.

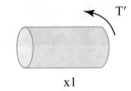

Figure 2.25: Internal torque, T'.

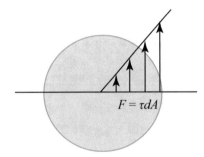

Figure 2.26: Shear stress distribution due to applied torque, T.

$$P = \sigma \int dA$$

$$\sigma = \frac{P}{A}. \tag{2.7}$$

2.5.2   APPLIED TORQUE

For an applied torque, T, as shown in Figure 2.25, the internal shear stress distribution, T', is assumed to be linear and illustrated in Figure 2.26. To derive Equation (2.8), we know the torque is comprised of two components. The force is τdA and moment arm is ρ:

$$\circlearrowleft + \sum T = 0$$

$$T - F\rho = 0$$

$$T - \tau\rho dA = 0$$

$$\tau = \frac{\tau_{\max}}{R}\rho \quad \text{where} \quad R = \text{radius}$$

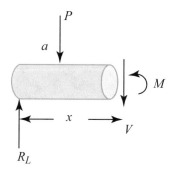

Figure 2.27: FBD for beam.

substituting for τ yields:

$$T - \frac{\tau_{\max}}{R}\rho^2 dA = 0$$

$$T = \frac{\tau_{\max}}{R}\int \rho^2 dA.$$

From Statics we know $\int \rho^2 dA$ is defined as the polar moment of inertia, J. Solving for $\tau_{\max}$ yields Equation (2.8):

$$\tau_{\max} = \frac{TR}{J}. \tag{2.8}$$

2.5.3   BENDING

Figure 2.10 showed a circular shaft with an applied load. This load will cause the shaft to bend, thus, inducing an internal bending stress. Figure 2.27 shows an FBD for a cut beam. From this the bending stress, σ_b, and shear stress, τ, will be derived. Figure 2.28 illustrates the internal bending stress distribution is linear with the maximum stress occurring at the top or bottom fiber. This is important to understand in Chapter 3 when the shaft diameter equations are developed under cyclic loading.

To derive the bending stress equation we first must find the neutral axis for the beam's cross-section. For a shaft the neutral axis it at its center. Defining upward as the positive y-direction it should be noted that the top fibers are in compression and the bottom fibers are in tension. It is not too important to memorize this since the derivation for the bending stress will indicate the positive and negative stress:

$$\circlearrowleft + \sum M = 0$$

$$M + P(x - a) = 0$$

$$M = -P(x - a)$$

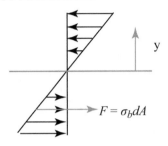

Figure 2.28: Internal stress distribution in beams.

define M as:

$$M = \sigma_b\, y\, dA$$

using the linear assumption σ_b is:

$$\sigma_b = \frac{\sigma_{\max}}{R} y \quad \text{where} \quad R = \text{radius}$$

substituing yields:

$$\frac{\sigma_{\max}}{R} y^2 dA = -P(x - a)$$

$$\frac{\sigma_{\max}}{R} \int y^2 dA = -P(x - a).$$

From Statics we know $\int y^2 dA$ is defined as the moment of inertia, I. Solving for $\sigma_{\max}$ yields Equation (2.9):

$$\sigma_{\max} = \frac{-P(x - a)}{I} R. \tag{2.9}$$

It should be obvious that this derivation is a special case with one load, P, located at a distance a from the left support. In general, from a course on stress analysis, the bending stress is given in Equation (2.10). As discussed earlier, the negative sign indicate that the top fibers are in compression:

$$\sigma_b = -\frac{Mc}{I}. \tag{2.10}$$

Figure 2.29 is the shear stress distribution for a shaft. It is included for completeness and left to the students to understand its derivation. Our focus is only on the outer fibers of the beam where shear stress, $\tau = 0$. Equation (2.11) can be found in any elementary text on stress analysis:

$$\tau = \frac{VQ}{It}. \tag{2.11}$$

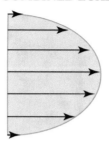

Figure 2.29: Shear stress distribution for bending.

Figure 2.30: Simple beam under axial load.

2.6   COMBINED LOADS: A MOTIVATIONAL EXAMPLE

Section 2.5 illustrated the internal stress distribution for axial, shear, and bending loads. This section demonstrates how to use Mohr's circle when a shaft is under combined loading to calculate the principal stresses. We start with Figure 2.30 which is a simple beam with a $1'' \times 1''$ cross-section with a 1000.0-lb axial load. Figure 2.31 is the FBD for the beam and we can solve for the axial stress induced in the beam using (2.7):

$$\sigma = \frac{P}{A}$$

$$\sigma = \frac{1000.0}{1.0} = 1000.0 \text{ psi.}$$

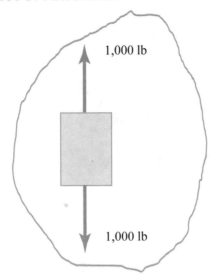

Figure 2.31: **FBD for beam.**

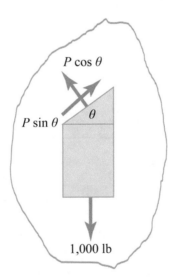

Figure 2.32: Slice along angle θ.

Now, let's cut the bar at an angle θ and draw the normal and tangential force seen shown in Figure 2.32. The normal stress was defined in Equation (2.7). In general the shear stress, τ, is given in Equation (2.12):

$$\tau = \frac{P_{tangential}}{A}. \tag{2.12}$$

For this example, the normal and shear stresses are defined as follows:

$$\sigma = \frac{P_{normal}}{A_\theta}$$

$$\tau = \frac{P_{tangential}}{A_\theta}$$

substituting

$$A_\theta = \frac{1}{\cos\theta}$$

$$P_{normal} = P\cos\theta$$

$$P_{tangential} = P\sin\theta$$

$$\sigma = P\cos^2\theta$$

$$\tau = P\cos\theta\sin\theta.$$

In keeping with the computational modeling approach, a simple program was written to plot Mohr's circle. The code is given below and the output is plotted in Figure 2.33.

Set Vector to angle Theta

Mohr's Circle

```
Theta=(0:360);
Width=input('Beam Width = ');
Depth=input('Beam Depth = ');
Load=input('Applied Load = ');
Area=Width*Depth;
stressSIGMA=Load*cosd(Theta)*cosd(Theta)/Area;
stresstau=Load*cosd(Theta)*sind(Theta)/Area;
scatter(stressSIGMA,stresstau);
```

From this example, the maximum stress of 1000.0 psi occurred when $\theta = 0$. The shear stress $\tau = 0$ at this point. The shear stress being equal to 0 is not by luck but illustrates that σ is a principal stress. For now recall, from strength of materials, that σ_1 and σ_2 occur when the shear stress $\tau = 0$. This was a simple example, however, in general the equations on any plane starts with the infinitesimal element shown in Figure 2.34.

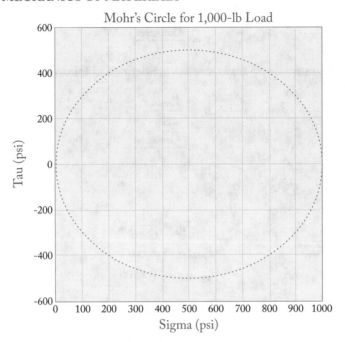

Figure 2.33: Mohr's circle.

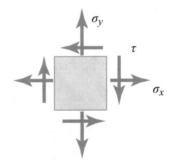

Figure 2.34: Infinitesimal element (positive orientation).

Figure 2.35 cuts an arbitrary plane at angle θ and Figure 2.36 shows the stresses acting in the normal and tangential directions. Using $\sum F_{x'} = 0$ and $\sum F_{y'} = 0$ leads to the development of Equations (2.13) and (2.14):

$$\sum F_{x'} = 0$$

$$\sigma_\theta dA - \sigma_x dA \cos\theta \cos\theta - \sigma_y dA \sin\theta \sin\theta + \tau dA \cos\theta \sin\theta + \tau dA \sin\theta \cos\theta$$

$$\sigma_\theta = \sigma_x \cos^2\theta + \sigma_y \sin^2\theta - 2\tau \cos\theta \sin\theta \tag{2.13}$$

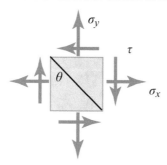

Figure 2.35: Cut along plane.

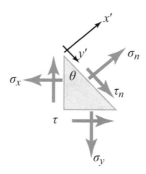

Figure 2.36: Element cut at angle θ.

$$\sum F_{y'} = 0$$

$$\tau_\theta dA - \sigma_x dA \cos\theta \sin\theta + \sigma_y dA \sin\theta \cos\theta - \tau dA \cos\theta \cos\theta + \tau dA \sin\theta \sin\theta$$

$$\tau_\theta = \sigma_x \cos\theta \sin\theta - \sigma_y \sin\theta \cos\theta + \tau \cos^2\theta - \tau \sin^2\theta. \tag{2.14}$$

Using the previous example, the infinitesimal element is drawn in Figure 2.37. Choosing $\theta = 28°$ yields $\sigma = P\cos^2\theta = 779.6$ psi and $\tau = P\cos\theta\sin\theta = 414.5$ psi. Using Equations (2.13) and (2.14) (remember now $\theta = 118°$) yields $\sigma_\theta = \sigma_y \sin^2\theta = 779.6$ psi and $\tau_\theta = -\sigma_y \sin\theta\cos\theta = 414.5$ psi.

Example 2.7 (Shaft under combined loads)
Figure 2.38 is a simply supported shaft under axial, torsional, and bending loads. It is best to solve this type of problem in steps.

Step 1: Draw the complete FBD similar to Figure 2.39.

Step 2: Calculate the axial stress:

$$\sigma = \frac{P}{A}$$

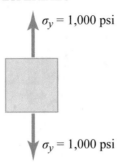

Figure 2.37: Example infinitesimal element set-up.

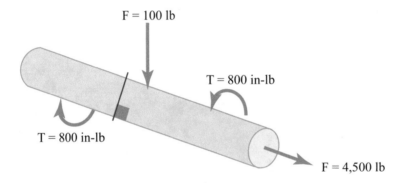

Figure 2.38: Shaft under multiple loads.

Table 2.1: Shaft data

Properties	Data
Length	20 in
Diameter	0.75 in
Yield	36,000 psi
Modulus	30 e 06

$$A = \frac{\pi D^2}{4} = \frac{\pi (0.75)^2}{4} = 0.442$$

$$\sigma = \frac{4500.0}{0.442} = 10186 \text{ psi.}$$

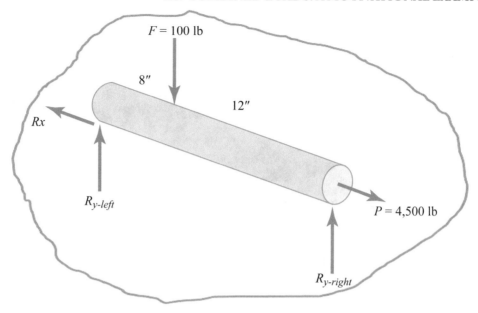

Figure 2.39: FBD for shaft.

Step 3: Calculate the shear stress from torque, T:

$$\tau = \frac{TR}{J}$$

$$J = \frac{\pi D^4}{32} = \frac{\pi (0.75)^4}{32} = 0.0311$$

$$\tau = \frac{(800.0)(0.375)}{0.0.0311} = 9647 \text{ psi.}$$

Step 4: Draw Shear and Bending Moment diagrams (Figure 2.40) given load, F.

Step 5: Calculate Bending Stress:

$$\sigma_b = \frac{MR}{I}$$

$$I = \frac{\pi D^4}{64} = \frac{\pi (0.75)^4}{64} = 0.01553.$$

Step 6: Set-up the infinitesimal element (Figures 2.41 and 2.42).

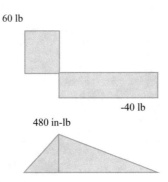

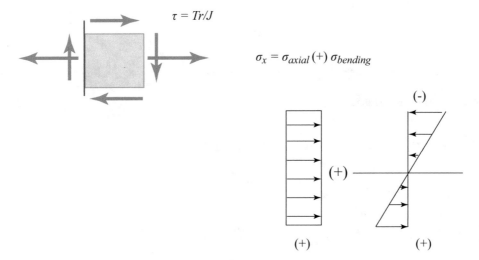

Figure 2.40: Shear and bending moment diagram.

$$\tau = Tr/J$$

$$\sigma_x = \sigma_{axial} (+) \sigma_{bending}$$

Figure 2.41: Example 2.4 infinitesimal element.

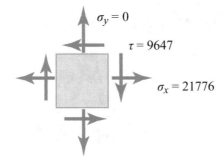

Figure 2.42: Element stresses.

Code for Equations (2.13) **and** (2.14)

Infinitesimal Element Derivaton Model

```
AngleTHETA=0:360;
sigmaX=input('sigmaX = ');
sigmaY=input('sigmaY = ');
tau=input('tau = ');
Create the Cos and Sin Matrix
MatrixCOS=cosd(AngleTHETA);
MatrixSIN=sind(AngleTHETA);
cosSQUARE=MatrixCOS.*MatrixCOS;
sinSQUARE=MatrixSIN.*MatrixSIN;
cossinMatrix=MatrixCOS.*MatrixSIN;
Calculate the normal and tangential stress of the element
sigmaTHETA= sigmaX*cosSQUARE + sigmaY*sinSQUARE
            -2*tau*cossinMatrix;
tauTHETA= sigmaX*cossinMatrix - sigmaY*cossinMatrix
            + tau*cosSQUARE - tau*sinSQUARE;
Find Principal Stresses
sigma1=max(sigmaTHETA);
sigma2=min(sigmaTHETA);
shearMAX=max(tauTHETA);
plot sigma vs. tau
scatter(sigmaTHETA,tauTHETA);
hold on
xline=[sigmaX,sigmaY];
yline=[tau,-tau];
line(xline,yline,'Color','red','LineStyle','--');
hold off;
```

Step 7: Analyze the results:

Figure 2.43 is the tau vs. sigma plot. Results:

$\sigma_1 = 25435$ psi

$\sigma_2 = -3659$ psi

$|\tau| = 15547$ psi.

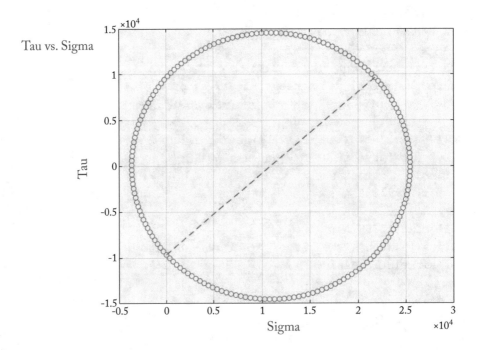

Figure 2.43: Tau vs. sigma.

CHAPTER 3

Shaft Sizing and Fatigue

3.1 INTRODUCTION

This chapter discusses two important points. The first is to show how the shaft diameter equation is derived. Technology students are often involved in design and performing preliminary calculations and it's imperative they understand the genesis of equations. Second, for students to understand the assumptions and limitations of the formula being used. In this chapter we will limit the equation for calculating the shaft diameter using the Soderberg fatigue curve under a constant torque.

3.2 DERIVATION OF SHAFT DIAMETER

The Soderberg fatigue curve is defined in Equation (3.1):

$$\frac{\sigma_a}{\sigma_e} + \frac{\sigma_m}{\sigma_y} = 1. \tag{3.1}$$

Before we develop the equations to find the shaft diameter incorporating the Soderberg's criterion, we must identify the equations for a static analysis. To start, let's assume we know the torque, T, and bending moment, M, applied to the shaft. In addition, we assume the shaft diameter, d, is known. We solve for the maximum bending stress using Equation (3.2) and Equations (3.3) and (3.4) solve for c and the moment of inertia, I, respectively. Next, the shear stress is defined in Equation (3.5) where the polar moment of inertia, J, is calculated using Equation (3.6). If only a load, that induces a bending moment, or a torque acts alone then we can use either Equation (3.7) or (3.8) to determine if our design is acceptable:

$$\sigma_b = \frac{Mc}{I} \tag{3.2}$$

$$c = \frac{d}{2} \tag{3.3}$$

$$I = \frac{\pi d^4}{64} \tag{3.4}$$

$$\tau = \frac{Tr}{J} \tag{3.5}$$

$$J = \frac{\pi d^4}{32} \tag{3.6}$$

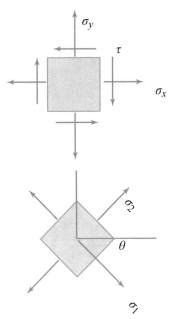

Figure 3.1: Infintesimal element.

$$\sigma_b \leq \frac{\sigma_y}{N} \tag{3.7}$$

$$\tau \leq \frac{\tau_y}{N}. \tag{3.8}$$

Typically, bending and shear stresses act simultaneously on the shaft. When loads act simultaneously, we can use Mohr's circle to find the principal stresses on the shaft. Figure 3.1 shows the infinitesimal element required to use the Mohr's circle process and Figure 3.2 draws Mohr's circle. In Chapter 2, we defined the principal stresses as the point on Mohr's circle where the shear stress is equal to zero. Mathematically, they are defined by Equations (3.9), (3.10), and (3.11) knowing that for shaft design $\sigma_y = 0$ and N is the factor of safety:

$$\sigma_1 = \frac{\sigma_x}{2} + \tau_{\max} \tag{3.9}$$

$$\sigma_2 = \frac{\sigma_x}{2} - \tau_{\max} \tag{3.10}$$

$$\tau_{\max} = \sqrt{\left(\frac{\sigma_x - \sigma_y}{2}\right)^2 + \tau^2}. \tag{3.11}$$

Equation (3.12) defines the von-Mises stress. We will use the von-Mises criterion to find the combined stress acting on the shaft and compare it to the allowable stress expressed in Equa-

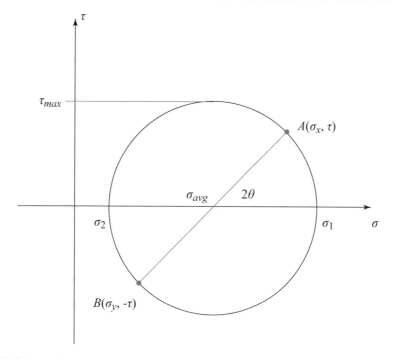

Figure 3.2: Mohr's circle.

tion (3.13):

$$\sigma_v = \sqrt{\sigma_1^2 - \sigma_1\sigma_2 + \sigma_2^2} \tag{3.12}$$

$$\sigma_v = \sqrt{\sigma_1^2 - \sigma_1\sigma_2 + \sigma_2^2} \le \frac{\sigma_y}{N}. \tag{3.13}$$

Now, we will substitute Equations (3.9), (3.10), and (3.11) into Equation (3.12). After some algebraic manipulation the von-Mises criterion is given in Equation (3.14) for σ_x and $\tau_{\max}$:

$$\sigma_v = \sqrt{\left(\frac{\sigma_x}{2} + \tau_{\max}\right)^2 - \left(\frac{\sigma_x}{2} + \tau_{\max}\right)\left(\frac{\sigma_x}{2} - \tau_{\max}\right) + \left(\frac{\sigma_x}{2} - \tau_{\max}\right)^2}$$

$$\sigma_v = \sqrt{\left(\frac{\sigma_x}{2}\right)^2 + 2\left(\frac{\sigma_x}{2}\right)\tau_{\max} + \tau_{\max}^2 - \left(\left(\frac{\sigma_x}{2}\right)^2 - \tau_{\max}^2\right) + \left(\frac{\sigma_x}{2}\right)^2 - 2\left(\frac{\sigma_x}{2}\right)\tau_{\max} + \tau_{\max}^2}$$

$$\sigma_v = \sqrt{\left(\frac{\sigma_x}{2}\right)^2 + 3\tau_{\max}^2}. \tag{3.14}$$

Substituting $\tau_{\max}$ (Equation (3.11)) into Equation (3.14) and some algebraic manipulation yields the von Mises formula, Equation (3.15), for a shaft under static conditions subjected to bending and shear stresses:

$$\sigma_v = \sqrt{\left(\frac{\sigma_x}{2}\right)^2 + 3\left(\sqrt{\left(\frac{\sigma_x}{2}\right)^2 + \tau^2}\right)^2}$$

$$\sigma_v = \sqrt{\left(\frac{\sigma_x}{2}\right)^2 + 3\left(\left(\frac{\sigma_x}{2}\right)^2 + \tau^2\right)}$$

$$\sigma_v = \sqrt{(\sigma_x)^2 + 3\tau^2}. \tag{3.15}$$

Let's return to Equation (3.1) and set up the shaft design equations for fatigue. Students and designers should keep in mind that failure under cyclic loading can occur below the yield strength of a material. We will make a few assumptions in this section.

1. Our shaft is only subjected to bending and torsion. Under certain conditions we can account for axial loads.

2. The torque is constant, $T_a = 0$.

3. The mean bending stress, $\sigma_m = 0$.

The Soderberg criterion is very conservative and accounts that failure could not occur by exceeding the material's yield strength. Simply, failure is caused by cyclic loading. Equation (3.1) defines σ_a as the amplitude stress given in Equation (3.16). Equation (3.17) defines the mean stress, σ_m:

$$\sigma_a = \frac{\sigma_{\max} - \sigma_{\min}}{2} \tag{3.16}$$

$$\sigma_m = \frac{\sigma_{\max} + \sigma_{\min}}{2}. \tag{3.17}$$

The endurance stress, σ_e, will be defined shortly and the yield stress, σ_y, is a material property. Equation (3.1) is multiplied by σ_y resulting in Equation (3.18). We do the same for the shear stress resulting in Equation (3.19):

$$\sigma_y \left(\frac{\sigma_a}{\sigma_e} + \frac{\sigma_m}{\sigma_y} = 1\right)$$

$$\frac{\sigma_y \sigma_a}{\sigma_e} + \sigma_m = \sigma_y. \tag{3.18}$$

From a design point of view we know $\sigma_x \leq \sigma_y$ or $(\sigma_x \leq \frac{\sigma_y}{N})$. Thus, $\sigma_x \leq \frac{\sigma_y \sigma_a}{\sigma_e} + \sigma_m$. Doing the same for the shear stress and substituting Equations (3.18) and (3.19) into Equation (3.15) yields Equation (3.20):

$$\tau_y \left(\frac{\tau_a}{\tau_e} + \frac{\tau_m}{\tau_y} = 1\right)$$

$$\frac{\tau_y \tau_a}{\tau_e} + \tau_m = \tau_y \tag{3.19}$$

$$\sigma_v = \sqrt{\left(\frac{\sigma_y \sigma_a}{\sigma_e} + \sigma_m\right)^2 + 3\left(\frac{\tau_y \tau_a}{\tau_e} + \tau_m\right)^2}. \tag{3.20}$$

Let's derive the mean and amplitude bending stress:

$$\sigma_{b\,max} = \frac{M_{max}c}{I} = \frac{M_{max}\frac{d}{2}}{\frac{\pi d^4}{64}} = \frac{32 M_{max}}{\pi d^3}.$$

Repeating the process to find the maximum shear stress due to applied torque, T, yields:

$$\tau_{max} = \frac{T_{max}r}{J} = \frac{T_{max}\frac{d}{2}}{\frac{\pi d^4}{32}} = \frac{16 T_{max}}{\pi d^3}.$$

For minimum bending and shear stress we just substitute M_{min} and T_{min}, respectively,

$$\sigma_{b\,min} = \frac{32 M_{min}}{\pi d^3}$$

$$\tau_{min} = \frac{16 T_{min}}{\pi d^3}.$$

Substituting the calculated minimum and maximum bending and shear stresses into Equations (3.16) and (3.17), respectively, yields the amplitude and mean stresses:

$$\sigma_a = \frac{\frac{32 M_{max}}{\pi d^3} - \frac{32 M_{min}}{\pi d^3}}{2}$$

$$\sigma_a = \frac{16}{\pi d^3}\left(M_{max} - M_{min}\right)$$

$$\sigma_m = \frac{16}{\pi d^3}\left(M_{max} + M_{min}\right)$$

$$\tau_a = \frac{8}{\pi d^3}\left(T_{max} - T_{min}\right)$$

$$\tau_m = \frac{8}{\pi d^3}\left(T_{max} + T_{min}\right).$$

For our case, knowing from mechanics of materials $\sigma_m = 0$ because the infinitesimal element at the top of the circular shaft initially has a compressive stress. After the shaft rotates 180° the element is under tension with the same magnitude ($M_{max} = -M_{min} = M$). Knowing the Torque, T, is a constant results in $T_{max} = T_{min} = T$. Therefore, $\tau_a = 0$. Our equations become:

$$\sigma_a = \frac{32M}{\pi d^3}$$

$$\sigma_m = 0$$

$$\tau_a = 0$$

$$\tau_m = \frac{16T}{\pi d^3}.$$

Now, we substitute into Equation (3.20) and perform some simple algebra yielding Equation (3.21):

$$\sigma_v = \sqrt{\left(\frac{\sigma_y \frac{32M}{\pi d^3}}{\sigma_e}\right)^2 + 3\left(\frac{16T}{\pi d^3}\right)^2}$$

$$\sigma_v = \sqrt{\left(\frac{\sigma_y}{\sigma_e}\right)^2 \left(\frac{32M}{\pi d^3}\right)^2 + 3\left(\frac{32T}{2\pi d^3}\right)^2}$$

$$\sigma_v = \sqrt{\left(\frac{\sigma_y}{\sigma_e}\right)^2 \left(\frac{32M}{\pi d^3}\right)^2 + \frac{3}{4}\left(\frac{32T}{\pi d^3}\right)^2}$$

$$\sigma_v = \sqrt{\left(\frac{\sigma_y}{\sigma_e}M\right)^2 \left(\frac{32}{\pi d^3}\right)^2 + \frac{3}{4}\left(\frac{32}{\pi d^3}\right)^2 T^2}$$

$$\sigma_v = \left(\frac{32}{\pi d^3}\right)\sqrt{\left(\frac{\sigma_y}{\sigma_e}M\right)^2 + \frac{3}{4}T^2}. \tag{3.21}$$

Finally, knowing that the von Mises stress, σ_v, is the design stress and must be $\leq \frac{\sigma_y}{N}$ we can now solve for the shaft diameter, d (Equation (3.22)):

$$\left(\frac{32}{\pi d^3}\right)\sqrt{\left(\frac{\sigma_y}{\sigma_e}M\right)^2 + \frac{3}{4}T^2} \leq \frac{\sigma_y}{N}$$

$$d^3 \geq \frac{32N}{\pi \sigma_y}\sqrt{\left(\frac{\sigma_y}{\sigma_e}M\right)^2 + \frac{3}{4}T^2}$$

$$d \geq \left(\frac{32N}{\pi \sigma_y}\sqrt{\left(\frac{\sigma_y}{\sigma_e}M\right)^2 + \frac{3}{4}T^2}\right)^{\frac{1}{3}}. \tag{3.22}$$

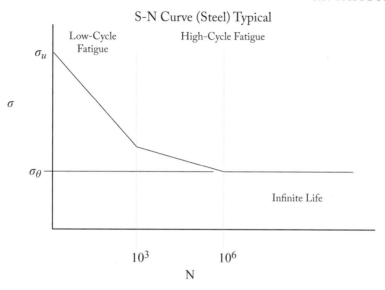

Figure 3.3: S–N curve.

3.3   FATIGUE CURVES

Figure 3.3 shows the typical S–N curve for steel. We now can discuss, σ_e which we defined earlier as the endurance stress. The endurance stress is defined as the maximum stress on a structure which can undergo infinite cylical cycles without failure due to fatigue. There are many curves that define infinite life. Figure 3.4 is the Soderberg infinite life fatigue curve and the combination of (σ_m, σ_a) must be below its infinite life curve.

For steel the endurance stress for steel is defined as:

$$\sigma'_e = 0.5\sigma_u \text{ when } \sigma_u \leq 100,000 \text{ psi}$$
$$\sigma'_e = 100,000 \text{ psi when } \sigma_u > 100,000 \text{ psi.}$$

The endurance stress defined above is just a starting point and must account for operating conditions. Therefore, we define the adjusted endurance using Equation (3.23)

$$\sigma_e = \sigma'_e \left(k_a k_b k_c k_d k_e k_f\right) \frac{1}{K_f}. \tag{3.23}$$

Factor k_a is the surface finish factor. Equation (3.24) is used to calculate k_a along with the coefficients for a and b given in Table 3.1:

$$k_a = a\sigma_u^b. \tag{3.24}$$

Factor k_b is the shaft diameter factor calculated using Equation (3.25). This is a Catch-22 situation. During the design process, the goal is to size the shaft diameter. However, k_b adjust the

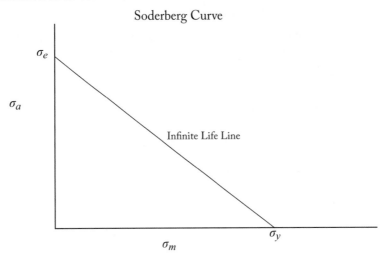

Figure 3.4: Soderberg fatigue curve.

Table 3.1: Surface factors

Surface Type	a(kpsi)	b(kpsi)
Ground	1.34	-0.85
Machined	2.7	-.27
Cold Drawn	2.7	-.27
Hot Roll	14.4	-.72
Forged	39.9	-.99

endurance stress, σ_e to account for the diameter. This becomes an iterative process. Simply, if our value for k_b is within range we are fine. If not, we must recalculate the diameter using the adjusted k_b factor. Equation (3.25) only applies to fatigue loading under bending and torsion. For axial fatigue loading $k_b = 1.0$:

$$k_b = 0.879d^{-0.107} \quad (0.11 < d < 2)$$
$$k_b = 0.91d^{-0.157} \quad (2 \leq d \leq 10). \tag{3.25}$$

The loading factor, k_c, is dependent on the type of load being applied to a structure. Equation (3.26) gives the coefficient for k_c as a function of load type. Factor k_d is dependent of the temperature (°F) and the formula is given in Equation (3.27). The loading factor, k_e, is reliability

Table 3.2: Reliability factors

Reliability (%)	Za	ke
50	0	1.0
90	1.288	0.897
95	1.645	0.868
99	2.326	0.814
99.9	3.091	0.753
99.99	3.791	0.702

factor and expressed in Equation (3.28). Table 3.2 gives the value for k_e for reliability value:

$$k_c = 1.0 \quad \text{(bending)}$$
$$k_c = 0.850 \quad \text{(axial)} \quad\quad\quad (3.26)$$
$$k_c = 0.59 \quad \text{(torsion)}$$

$$k_d = 0.975 + 0.432\left(10^{-3}\right)T - 0.115\left(10^{-5}\right)T^2 + 0.104\left(10^{-8}\right) - 0.595\left(10^{-12}\right)T^4 \quad (3.27)$$

$$k_e = 1 - 0.8Z_a \quad\quad\quad (3.28)$$

$$Z_a = \frac{x - \bar{x}}{\sigma_x}.$$

Finally, factor k_f, accounts for miscellaneous effects, such as, corrosion, electrolytic plating, metal spraying, cyclic frequency, and frettage corrosion.

Students should recall from mechanics of material stress concentration factors. The simple example will illustrate how to obtain K_f for a plate with a hole:

$$K_f = \frac{maximum\ stress\ in\ notched\ specimen}{stress\ in\ notch\ free\ specimen}$$

K_t is the stress concentration factor under static loading. We adjust this under fatigue loading by solving Equations (3.29) and (3.30):

$$q = \frac{K_f - 1}{K_t - 1} \quad\quad\quad (3.29)$$

$$K_f = q(K_t - 1) + 1. \quad\quad\quad (3.30)$$

Example 3.1   (Fatigue Stress factor K_f)
The plate shown in Figure 3.5 has a stress $\sigma_x = 10{,}000$ psi. Given $d/w = 0.4$ from Figure 3.6

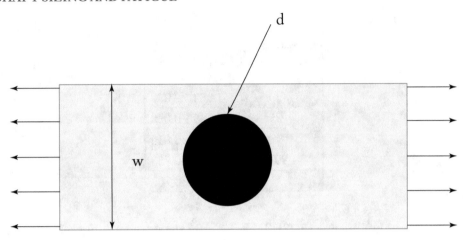

Figure 3.5: Thin plate.

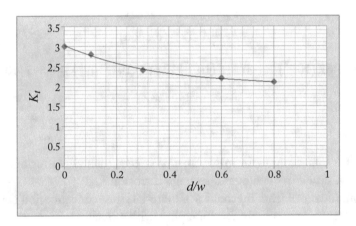

Figure 3.6: Static stress concentration factor.

we find $K_t = 2.3$. Remembering from strength of materials $K_t = \frac{\sigma_{max}}{\sigma_o}$ we can solve for the maximum stress in the thin plate under static loading. $\sigma_{max} = (2.3)(10{,}000) = 23{,}000$ psi.

Example 3.2 (Endurance Stress)

Given we are to design shaft made of steel wih $\sigma_u = 100{,}000$ psi. Given $q = 0.9$ and $K_t = 2.3$ we can solve for K_f using Equation (3.30). Use fatigue factors shown in Table 3.3:

$$K_f = q(K_t - 1) + 1 = 0.9(2.3 - 1) + 1 = 2.17$$

Table 3.3: Fatigue factors

Factor	Value
a	0.8
b	0.95
c	1.0
d	0.9
e	0.814
f	0.8

$$\sigma_e = \left(\frac{1}{2}\sigma_u\right) k_a k_b k_c k_d k_e k_f \frac{1}{K_f}$$

$$= \frac{1}{2}(100{,}000)(0.9)(0.95)(1.0)(0.9)(0.814)(0.8)\left(\frac{1}{2.17}\right)$$

$$= 11546 \text{ psi.}$$

CHAPTER 4

Machine Components

4.1   INTRODUCTION

This chapter covers the fundamentals of machine components. We limit our topics to:

1. Power equations

2. Spur gears

3. Straight bevel gears

4. Flat belts

5. Chains

6. Bearings

7. Clutches and Springs (A relationship)

8. Keyways

and use a systems approach to illustrate how loads and torques on various machine components are transferred to the shaft.

4.2   POWER EQUATIONS

It is critical for students to understand the horsepower (HP) equations. In particular, the care required for units. Equation (4.1) states the HP in terms of the speed, (rpm) and torque, T (in-lb). Equation (4.2) defines the tangential force, F_t, (pounds) in terms of diameter, d (inches). Finally, Equation (4.3) gives an alternate equation for HP in terms of velocity, v (ft/min). These equations are used extensively to calculate the forces and torques imparted by gears, belts, and chains onto shafts:

$$\text{HP} = \frac{Tn}{63,000} \tag{4.1}$$

$$T = F_t \frac{d}{2} \tag{4.2}$$

$$\text{HP} = \frac{F_t V}{33,000}. \tag{4.3}$$

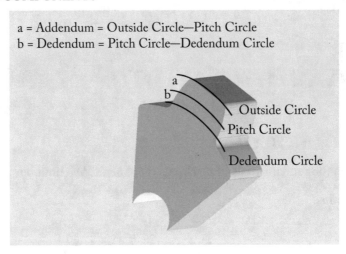

a = Addendum = Outside Circle—Pitch Circle
b = Dedendum = Pitch Circle—Dedendum Circle

Figure 4.1: Spur gear nomenclature.

4.3   SPUR GEARS

Figure 4.1 illustrates the nomenclature that is required for students to understand for spur gears. Study it carefully and become familiar with the terms. The most important term is the pitch diameter, d, or simply the gear's diameter. Loads applied to gears and the loads imparted onto a shaft by spur gears all take place from the gear's pitch circle. In reality the pitch circle does not exists, but when performing engineering calculations we need a frame of reference. Equation (4.4) defines the diametral pitch, P_d. The diametral pitch is the number of teeth per inch of the gear. Students often confuse the diametral pitch with pitch diameter. Just be conscience the two terms are not interchangeable and represent different gear characteristics. N is the number of teeth. Equations (4.5)–(4.7) defines the addendum, a, dedendum, b, and the face width, F, respectively. Equation (4.8) calculates the circular pitch which is defined as the distance from a point on the gear tooth to an identical point on the next tooth (measured along the pitch circle):

$$P_d = \frac{N}{d} \tag{4.4}$$

$$a = \frac{1}{P_d} \tag{4.5}$$

$$b = \frac{1.25}{P_d} \tag{4.6}$$

$$F = \frac{12}{P_d} \tag{4.7}$$

$$p = \frac{\pi d}{N} = \frac{\pi}{P_d}. \tag{4.8}$$

Example 4.1

Given a spur gear with $P_d = 10$ and $N = 35$, find the addendum, dedendum, face width, and gear diameter:

$$a = \frac{1}{10} = 0.1''$$

$$b = \frac{1.25}{10} = 0.125''$$

$$F = \frac{12}{10} = 1.2''$$

$$d = \frac{N}{P_d} = \frac{35}{10} = 3.5''.$$

The pitch circle is theoretically where gears engage and calculated using Equation (4.8):

$$p = \frac{\pi d}{N} = \frac{\pi}{P_d} = \frac{\pi}{10} = 0.314.$$

4.4   SPUR GEAR KINEMATICS

In the previous section we discussed individual gears. In this section, we introduce gear trains and methods to calculate velocity ratios and gear velocity. It is critical to understand how to build gear trains to achieve a desirable output speed or ratio. In Figure 4.2, a motor (shown in blue) has a gear attached to its shaft. Gears 2 and 3 share an intermediate shaft and gear 3 engages gear 4 to achieve the desired output speed. Figure 4.3 is the gear schematic.

Gear 1 (attached to the motor's shaft) engages gear 2. Since gears 2 and 3 share the same intermediate shaft, both gears must rotate at the same rotational velocity, ω. Finally, gear 3 engages output gear 4. At the point of engagement the velocity on both gears are identical. Equation (4.9) calculates the gear's velocity and Equation (4.10) calculates velocity in terms of ft/min. The velocity of gears 1 and 2 are identical at the point of engagement. Therefore, using the equality shown in Equation (4.11), we define the velocity ratio as the angular velocity, ω_{driver}, of the driver (typically attached to motor) divided by the angular velocity ω_{driven}. Equation (4.12) defines the velocity ratio, vr. **Note:** Do not confuse the velocity of a gear with its angular velocity. At the point of engagement, the velocity is the same the angular velocity is not:

$$V_{gear} = \omega r \tag{4.9}$$

$$\omega = 2\pi n$$

$$r = \frac{d}{2}$$

Figure 4.2: Sketch of gear train idea.

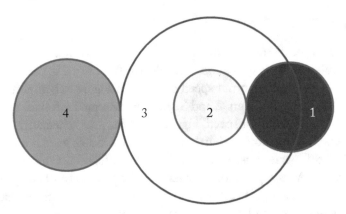

Figure 4.3: Gear train schematic.

$$V_{gear} = \left(\frac{2\pi \ \text{rad}}{\text{rev}} \right) \left(\frac{d \ \text{in}}{2} \right) \left(\frac{\text{ft}}{12 \ \text{in}} \right) \left(n \frac{\text{rev}}{\text{min}} \right)$$

$$V_{gear} = \frac{\pi d n}{12} \tag{4.10}$$

$$V_{gear1} = V_{gear2} = \omega_1 d_1 = \omega_2 d_2 \tag{4.11}$$

$$vr = \frac{\omega_{driver}}{\omega_{driven}} = \frac{\omega_1}{\omega_2}. \tag{4.12}$$

Table 4.1: Gear diameters

Gear	Diameter
1	9
2	15
3	5
4	8

Equation (4.13) rearranges Equation (4.11) to calculate the ratio of gear diameters to rotational velocity. Equation (4.14) shows the ratio for gear radius: Remembering the diametral pitch defined by Equation (4.4) and knowing P_d, must be the same for mating gears, we can use Equations (4.15)–(4.17) and find the ratio as a function of gear teeth. Equation (4.18) gives the velocity ratio, vr, as a function of gear teeth, diameter, or radius:

$$\frac{d_2}{d_1} = \frac{\omega_1}{\omega_2} \tag{4.13}$$

$$\frac{d_2}{d_1} = \frac{2r_2}{2r_1} = \frac{r_2}{r_1} = \frac{\omega_1}{\omega_2} \tag{4.14}$$

$$d_1 = \frac{N_1}{P_d} \tag{4.15}$$

$$d_2 = \frac{N_2}{P_d} \tag{4.16}$$

$$\frac{d_2}{d_1} = \frac{\dfrac{N_2}{P_d}}{\dfrac{N_1}{P_d}} = \frac{N_2}{N_1} \tag{4.17}$$

$$vr = \frac{\omega_1}{\omega_2} = \frac{d_2}{d_1} = \frac{r_2}{r_1} = \frac{N_2}{N_1}. \tag{4.18}$$

Example 4.2   (Gear Train)
Given the data in the Table 4.1 and $\omega_1 = 2{,}000$ rpm find the speed of each gear in the gear train:

$$\omega_2 = \omega_1 \frac{d_1}{d_2} = 2{,}000 \frac{9}{15} = 1200 \text{ rpm.}$$

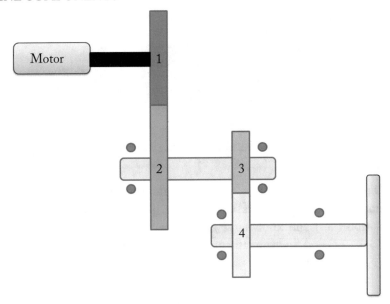

Figure 4.4: Gear train design.

We know that gear 3 is on the same shaft as gear 2. Thus, $\omega_3 = \omega_2 = 1200$ rpm. Finally, we can calculate the output gear:

$$\omega_4 = \omega_3 \left(\frac{d_3}{d_4} \right) = 1200 \left(\frac{5}{8} \right) = 750 \quad \text{rpm.}$$

4.5   SPUR GEAR STRESS ANALYSIS

This section gives an overview of calculating the stress in spur gears using the Lewis' and American Gear Manufacturer Association (AGMA) equations. From Figure 4.5 we see that spur gears are engaged at the pitch circle and assumed to act like a cantilever beam. Figure 4.6 is a schematic of how forces are imparted from the driver to the driven gears. It is critical for students to understand this sketch. The angle ϕ is gear's pressure angle (typically 14.5, 20, or 25°). The force, F, will be broken into two components: the tangential force, F_t, which transmits torque and the radial force, F_r, which is does not contribute to transmitting torque. Although the radial force does not contribute to transmitting torque, its load is critical when designing shafts and sizing bearings:

$$F_t = F \cos(\phi) \tag{4.19}$$

$$F_r = F \sin(\phi). \tag{4.20}$$

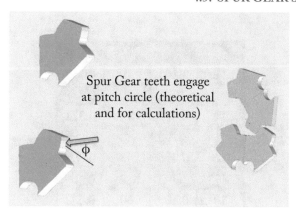

Figure 4.5: Engaged spur gears.

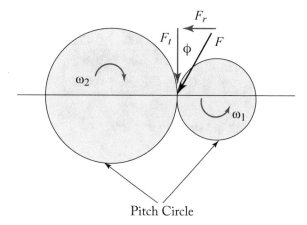

Figure 4.6: Gear kinematic sketch.

Solving Equations (4.19) and (4.20) for F yields Equation (4.21) to calculate directly F_r as a function of F_t:

$$F = \frac{F_t}{\cos(\phi)}$$

$$F = \frac{F_r}{\sin(\phi)}$$

$$\frac{F_t}{\cos(\phi)} = \frac{F_r}{\sin(\phi)} \tag{4.21}$$

$$F_r = F_t \frac{\sin(\phi)}{\cos(\phi)}$$

$$= F_t \tan(\phi).$$

Table 4.2: Form factor

N	Y	N	Y
12	0.245	28	0.353
13	0.261	30	0.359
14	0.277	34	0.371
15	0.290	38	0.384
16	0.296	43	0.398
17	0.303	50	0.409
18	0.309	60	0.422
19	0.314	75	0.435
20	0.322	100	0.447
21	0.328	150	0.460
22	0.331	300	0.472
24	0.337	400	0.480
26	0.346	Rack	0.485

The Lewis equation for calculating the stress in a gear is:

$$\sigma_b = \frac{F_t P_d}{bY} K_v, \tag{4.22}$$

where b is the face width and Y is the Lewis from factor given in Table 4.2 for a 20° pressure angle. K_v is the Barth dynmaic load factor and found using Equation (4.23) or (4.24). Finally, Equation (4.25) is the AGMA equation:

$$K_v = \frac{600 + V}{600} \tag{4.23}$$

$$K_v = \frac{1200 + V}{1200} \tag{4.24}$$

$$\sigma_b = \frac{F_t P_d}{bY} \frac{K_a K_b K_m K_s}{K_v} \tag{4.25}$$

K_a – *Application factor (1–2.75)*

K_b – *Rim thickness factor (non-solid gears)*

K_m – *Load distribution factor (1–2)*

K_s – *Size factor (1–1.4)*

K_v – *Dynamic factor (0.5–0.98).*

Example 4.3   (Gear Stress)

Given a 43 tooth gear, $P_d = 8$, and face width of 1/2″ determine the stress on the gear induced by a 4 HP, 1,000 rpm motor. Use the Lewis equation and Equation (4.24) to calculate the Barth velocity factor:

$$HP = \frac{Tn}{63,000}$$

$$T = \frac{63,000)(4)}{1,000} = 252 \text{ in} - \text{lb}$$

$$d = \frac{N}{P_d} = \frac{43}{8} = 5.375 \text{ in.}$$

$$F_t = \frac{2T}{d} = \frac{(2)(252)}{5.375} = 93.8 \text{ lb}$$

$$V = \frac{(\pi)dn}{12} = \frac{(\pi)(5.375)(1000)}{12} = 1407.2 \text{ ft/min}$$

$$K_v = \frac{1200 + V}{1200} = \frac{(1200 + 1407.2)}{1200} = 2.173$$

$$Y = 0.398 \quad \text{see Table 4.1}$$

$$\sigma_b = \frac{F_t P_d}{bY} K_v = \frac{(93.8)(8)}{(0.5)(0.398)}(2.173) = 8194 \text{ psi.}$$

If data are not available for allowable stress a rule of thumb is:

$$\sigma_b \leq \sigma_{allowable} = \frac{1}{3}\sigma_u.$$

Assume $\sigma_u = 60,000$ psi and check.

$$\sigma_b = 8194 \leq \frac{1}{3}(6,000) = 20,000 \text{ psi.}$$

4.6   STRAIGHT BEVEL GEARS

A disadvantage of spur gears is the design is limited to using parallel shafts. This could be an issue if space is limited in the gear box. Straight bevel gears overcome this issue but introduces an axial load on the shaft. Calculating the axial load is not difficult but needs to be taken into account when sizing the shaft's bearings. Figure 4.7 illustrates a straight bevel gear design. Figure 4.8 illustrates how the force transmitted by the pinion is broken into the axial and radial components.

From Figure 4.9, Equations (4.26) and (4.27) calculate the axial and tangential forces, respectively. Using Equation (4.30) and Figures 4.10 and 4.11, the power of Newton's Law of equal but opposite forces occurs on the pinion gear.

Figure 4.7: **Straight bevel gears.**

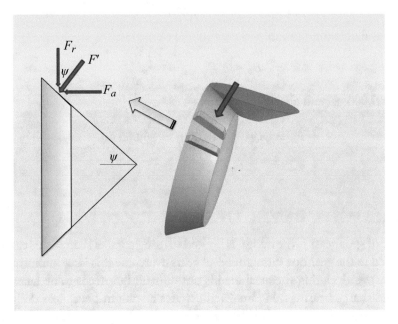

Figure 4.8: **Forces.**

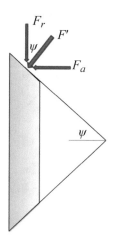

Figure 4.9: Radial and axial forces.

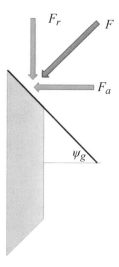

Figure 4.10: Forces on straight bevel-gear.

$$F_a = F_t \tan(\phi) \sin(\psi) \tag{4.26}$$

$$F_r = F_t \tan(\phi) \cos(\psi) \tag{4.27}$$

$$\Gamma = \psi_g + \psi_p \tag{4.28}$$

$$F_{a_{gear}} = F_{r_{pinion}} = F_t \tan(\phi) \sin(\psi_g) = F_t \tan(\phi) \cos(\psi_p) \tag{4.29}$$

$$F_{r_{gear}} = F_{a_{pinion}} = F_t \tan(\phi) \cos(\psi_g) = F_t \tan(\phi) \sin(\psi_p). \tag{4.30}$$

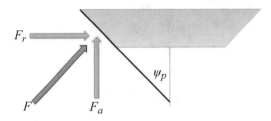

Figure 4.11: Forces on straight bevel-pinion.

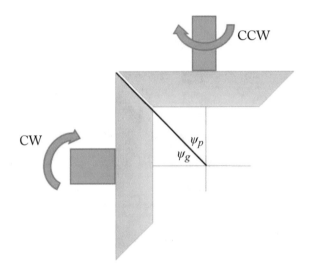

Figure 4.12: Example 4.4.

Example 4.4 (Straight Bevel Gears)
A pinion gear has the following data:

$$\Gamma = 90°$$
$$\psi_p = 22°$$
$$\phi = 20°$$
$$d_p = 10 \text{ in}$$
$$d_g = 25 \text{ in.}$$

Find the velocity ratio, axial and radial forces on gear and pinion, and torque on gear shaft. The pinion is attached to a 20 HP motor running at 500 rpm. See Figure 4.12.

$$\psi_g = \Gamma - \psi_p = 90. - 22. = 68°$$

$$V = \frac{\pi d n}{12} = \frac{\pi(10)(500)}{12} = 1309 \text{ ft/min}$$

$$HP = \frac{F_t V}{33,000}$$

$$F_t = \frac{(33,000)(20)}{1309} = 504.2 \text{ lb}$$

$$T_p = F_t \frac{d_p}{2} = (504.2)\frac{10}{2} = 2521 \text{ in-lb}$$

$$F_{a_{gear}} = F_t \tan(\phi) \sin(\psi_g) = (504.2)\tan(20)\sin(68) = 170.2 \text{ lb}$$

$$F_{r_{gear}} = F_t \tan(\phi) \cos(\psi_g) = (504.2)\tan(20)\cos(68) = 68.7 \text{ lb}$$

$$T_g = F_t \frac{d_g}{2} = (504.2)\frac{25}{2} = 6302.5 \text{ in-lb.}$$

Let's perform a simple check to show the axial force acting on the pinion is identical to the radial force acting on the gear. It is critical to understand radial, tangential, and axial loads are calculated along with the torques produced. These loads and torques are applied to the shaft and used during fatigue analysis to ensure the shaft is properly sized as well as properly choosing roller or ball bearings:

$$F_{a_{pinion}} = F_t \tan(\phi) \sin(\psi_p) = (504.2)\tan(20)\sin(22) = 68.7 \text{ lb}$$

$$F_{r_{pinion}} = F_t \tan(\phi) \cos(\psi_p) = (504.2)\tan(20)\cos(22) = 170.2 \text{ lb.}$$

4.7   FLAT BELTS

This section limits our discussion on belts to only flat belts. There are several advantages to use belt drives vs. gears:

Can transmit torques over long distance
Cheaper
Precise alignment not critical
Smoother and quieter
Durable

Figure 4.13 illustrates a simple flat belt design. We develop the design equations using the FBD illustrated in Figure 4.14. Equation (4.31) is the formula for centrifugal force. To find dF_c, we define m' as the mass per unit length in lbs/ft and develop Equation (4.32):

$$F_c = m a_n = m \frac{v^2}{r} \tag{4.31}$$

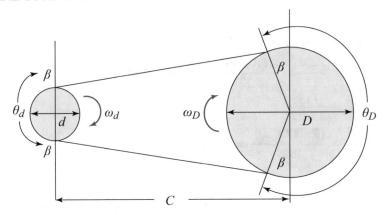

Figure 4.13: Flat belt schematic.

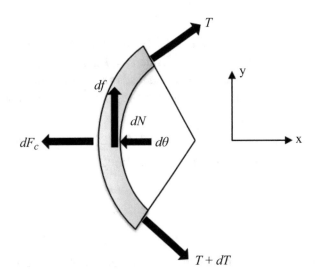

Figure 4.14: FBD belt section.

$$m = m'(rd\theta). \tag{4.32}$$

Given $v = \omega r$ and m, we can rewrite the centrifugal force equation as:

$$dF_c = m'(rd\theta)\frac{v^2}{r} = m'(rd\theta)v^2 = m'r^2\omega^2 d\theta. \tag{4.33}$$

Now, we return to the FBD for the flat belt and sum the forces in the x and y directions to develop the equations to calculate belt tensions. Belts have a tight side, depicted as $T + dT$ and a slack side, T:

$$\rightarrow + \sum F_x = 0 = -dF_c - dN + (T + dT)\sin\left(\frac{d\theta}{2}\right) + T\sin\left(\frac{d\theta}{2}\right) + df = 0. \quad (4.34)$$

$$\uparrow + \sum F_y = 0 = T\cos\left(\frac{d\theta}{2}\right) - (T + dT)\cos\left(\frac{d\theta}{2}\right) = 0. \quad (4.35)$$

Knowing:

$$\cos\left(\frac{d\theta}{2}\right) \cong 1$$

$$\sin\left(\frac{d\theta}{2}\right) \cong \frac{d\theta}{2}1$$

$$dT\sin\left(\frac{d\theta}{2}\right) \cong 0$$

we can rearrange Equations (4.34)–(4.35) and solve for tension T:

$$-dF_c - dN + Td\theta = 0$$
$$dN = Td\theta - dF_c$$
$$-dT + df = 0$$
$$dT = df = \mu dN = \mu(Td\theta - dF_c) = \mu Td\theta - \mu m' r^2 \omega^2 d\theta.$$

A little algebraic manipulation yields Equation (4.36). But, this is a first-order non-homogenous linear differential equation. Thus, Equation (4.37) yields the solution

$$\frac{dT}{d\theta} - \mu T = -\mu m' r^2 \omega^2 \qquad (4.36)$$

$$\frac{T_1 - F_c}{T_2 - F_c} = e^{c'}. \qquad (4.37)$$

Care must be taken in finding the coefficient, c'. It is the minimum of $(\mu_d \theta_d, \mu_D \theta_D)$. If the small and large pulleys are the same material then $c' = \mu_d \theta_d$, but this does not always true. To find c', perform the following:

$$\beta = \sin^{-1}\left(\frac{D - d}{2C}\right)$$
$$\theta_D = \pi + 2\beta$$
$$\theta_d = \pi - 2\beta$$
$$c' = \min(\mu_d \theta_d, \mu_D \theta_D).$$

Table 4.3: Belt data

Driver	Data
HP	12.5
n	1,500 rpm
d	4"
C	60"
vr	3
Belt	Data
Density	0.04 lb/in ^3
Coeff friction	0.6
b	4"
t	0.25"

Equation (4.3) is modified and shown in Equation (4.38). Equation (4.39) is the torque transmitted by flat belts onto its attached shaft. The centrifugal force, F_c, is calculated using Equations (4.40) and (4.41). Finally, calculate the velocity of the belt using Equation (4.10):

$$HP = \frac{(T_1 - T_2)V}{33,000} \tag{4.38}$$

$$T_{shaft} = T_1 \frac{d}{2} - T_2 \frac{d}{2}$$
$$T_1 - T_2 = \frac{2T_{shaft}}{d} \tag{4.39}$$

$$F_c = \frac{w'}{g} \left(\frac{V}{60}\right)^2 \tag{4.40}$$

$$w' = 12\rho bt. \tag{4.41}$$

Example 4.5 (Flat Belts)
Using the data provided in the table, find F_c, T_1, T_2, and determine if the stress induced in the belt is less than the allowable stress. Given that $\sigma_{all} = 1600$ psi.

$$vr = \frac{\omega_1}{\omega_2} = \frac{D}{d} = 3$$

$$D = 3d = (3)(4) = 12''$$

$$w' = 12\rho bt = 12(0.04)(4)(0.25) = 0.48 \text{ lb/ft}$$

$$V = \frac{\pi dn}{12} = \frac{\pi(4)(1500)}{12} = 1570.8 \text{ ft/min}$$

$$F_c = \frac{w'}{g}\left(\frac{V}{60}\right)^2 = \frac{0.48}{32.2}\left(\frac{(1570)}{60}\right)^2 = 10.22 \text{ lb}$$

$$\beta = \sin^{-1}\left(\frac{(D-d)}{2C}\right) = \sin^{-1}\left(\frac{8}{120}\right) = 0.0667 \text{ radians}$$

$$\theta = \pi - 2\beta = \pi - (2)(0.0667) = 3.0$$

$$c' = \mu\theta_d = (0.6)(3) = 1.8$$

$$e^{c'} = 6.05$$

$$\frac{(T_1 - F_c)}{(T_2 - F_c)} = e^{c'} = 6.05$$

$$\frac{(T_1 - 10.22)}{(T_2 - 10.22)} = 6.05$$

$$T_1 - 6.05T_2 = -51.6.$$

Using:

$$HP = \frac{(T_1 - T_2)V}{33,000}$$

$$15 = \frac{(T_1 - T_2)(1570.8)}{33,000}$$

$$(T_1 - T_2) = 262.6$$

Solving:

$$T_1 - 6.05T_2 = -51.6$$

$$T_1 - T_2 = 262.6$$

yields

$$T_1 = 324.8$$

$$T_2 = 62.2$$

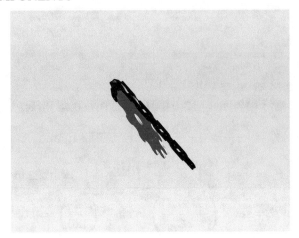

Figure 4.15: Sprocket and chain.

$$T_{shaft} = T_1 \frac{d}{2} - T_2 \frac{d}{2} = (324.8)(2) - (62.2)(2) = 525.2 \, \text{in-lb}$$

$$\sigma = \frac{T_1}{a_{belt}} = \frac{324.8}{1} = 324.8 \, \text{psi} \; < \sigma_{allowable}.$$

4.8 ROLLER CHAIN

This section limits our discussion to roller chains. Figure 4.15 illustrates a simple roller chain and sprocket. Chains are well established by code (ANSI/ASME Standard B29.1M-1993) and standards. Charts are readily available for chain's rated horsepower or strength. First, Equation (4.42) gives the chain velocity. Equation (4.43) modifies the horsepower equation in terms of chain pitch:

$$V_{Chain} = \frac{Npn}{12} \, \text{ft/min}, \qquad (4.42)$$

where
N – number of teeth on a sprocket (Note: Can be driver or driven),
p – chain pitch in inches,
n – speed in rpm.

$$\text{HP} = \frac{F_t V}{33,000} = \frac{F_t Npn}{396,000}. \qquad (4.43)$$

Equation (4.44) is the chain length in pitches where we define the variables as:

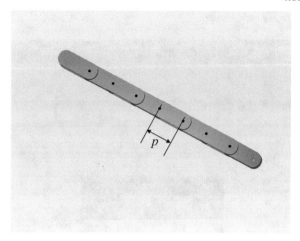

Figure 4.16: Illustration of pitch (p) definition.

N_1 – number of teeth on driver sprocket,
N_2 – number of teeth on driven sprocket,
C – center to center distance in inches,
p – chain pitch in inches,
n – speed in rpm.

$$\frac{L}{p} = \frac{2C}{p} + \frac{N_1 + N_2}{2} + \frac{p(N_2 - N_1)^2}{4\pi^2 C}. \tag{4.44}$$

For design purposes, Equation (4.45) defines the effective horsepower imparted on the chain. Equation (4.46) calculates the chain's horsepower capacity. Equating Equations (4.45) and (4.46) allows the designer to choose the appropriate chain:

HP_{nom} – Typically the "off-the-shelf" or data of motor being used.
K_d – Design factor.
K_s – Service factor (Table 4.4).
HP_r – Chain's rated horsepower (Obtained from code of manufacturer).
K_2 – Multiple chain strand factor (Table 4.5).

$$HP_{effective} = HP_{nom} K_d K_s \tag{4.45}$$

$$HP_{cap} = HP_r K_2 \tag{4.46}$$

$$HP_{nom} K_d K_s = HP_r K_2 \tag{4.47}$$

$$HP_r = \frac{HP_{nom} K_d K_s}{K_2}. \tag{4.48}$$

Table 4.4: Service factor K_s

Load	IC Engine Hydraulic	Electric Motor	IC Engine Mechanical
Smooth	1.0	1.0	1.2
Moderate shock	1.2	1.3	1.4
Heavy shock	1.4	1.5	1.7

Table 4.5: Strand factor K_2

Number of Chains	Factor
1	1.0
2	1.7
3	2.5
4	3.3
5	3.9
6	4.6
8	6.0

Example 4.6   (Roller Chain)

Given an electric motor at 25 HP, speed of 1000 rpm, and running under moderate shock conditions. Assuming one chain running on 17 tooth sprocket and ignoring the design factor, calculate the chain's velocity and rated horsepower:

$$K_s = 1.3$$
$$K_d = 1.0$$
$$K_2 = 1.0$$

$$\text{HP}_r = \frac{\text{HP}_{nom} K_d K_s}{K_2} = \frac{(25)(1.)(1.3)}{1.0} = 32.5 \text{ hp.}$$

We find that a chain with $p = 1$ is sufficient:

$$V_{Chain} = \frac{Npn}{12} = \frac{(17)(1)(1000)}{12} = 583.3 \text{ ft/min}$$

$$F_t = \frac{(32.5)(33,000)}{583.3} = 1838.7 \text{ lb.}$$

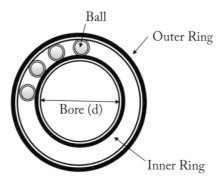

Figure 4.17: Bearing nomenclature.

4.9 BEARINGS

In this section we use the life formulas for bearings and show how the shaft's reaction loads are the radial loads acting on a bearing. Figure 4.17 illustrates a simple roller bearing. Equation (4.49) gives the life equations for bearings. Equation (4.49) is modifying to use catalog data by setting $L_2 = 1 \times 10^6$ revolutions and $P_2 = C$:

$$\frac{L_1}{L_2} = \left(\frac{P_2}{P_1} \right)^k \tag{4.49}$$

$$L_1 = \left(\frac{C}{P_1} \right)^k \quad L_2 = \left(\frac{C}{P_1} \right)^k (1 \times 10^6) \text{ revolutions,} \tag{4.50}$$

where $k = 3$ for ball bearings and $k = 3.33$ for cylindrical roller bearings. To calculate bearing life in hours use Equation (4.51) where n- rpm. Finally, Equation (4.52) calculates an equivalent radial load acting on the bearing when an axial load is present. This occurs when the desing is using a straight bevel gear:

$$L_{\text{Hrs}} = \frac{L_1}{60n} \tag{4.51}$$

$$P = VXR + YF_a, \tag{4.52}$$

where

  $V = 1.2$ for mount rotating or $V = 1.0$ if shaft rotates,
  R – Radial load,
  F_a – Axial load.

  Using Equation (4.53), the values for X and Y are obtained from manufacturer's data. The value C_o is the bearing's static load also obtained from the manufacturer:

$$e = \frac{F_a}{C_o}. \tag{4.53}$$

Example 4.7   (Bearing)
After designing a shaft the left reaction load (acting on the bearing) $= 1216$ lb. The right reaction load $= 800$ lb.

(a) Determine the life of the ball bearing given a bearing chosen from a catalog is rated for a radial load of 4240 lb.

(b) Given the shaft rotates at 200 rpm for 3 hours/day, how many days will the bearing last before it needs to be replaced.

(a)

$$L_1 = \left(\frac{C}{P_1}\right)^k (1 \times 10^6) = \left(\frac{4240}{1216}\right)^3 = 42.4 \times 10^6 \text{ revolutions}$$

(b)

$$L_{\text{Hrs}} = \frac{L_1}{60n} = \frac{42.4 \ 10^6}{(60)(200)} = 3534 \text{ hours}$$

$$L_{\text{Days}} = \frac{3534}{3} = 1178 \text{ days.}$$

Example 4.8   (Shaft/Bearing relationship)
This example illustrates how to apply the reaction forces to a bearing choice problem. Figure 4.18 shows a simple shaft supported by two ball bearings. At the left end, a flat belt and pulley are attached and impart a load of 1100 lb on the shaft as well as a torque of 1000 in-lb. After designing a shaft the left reaction load (acting on the bearing) $= 1216$ lb. A spur gear imparts a 700-lb and 1450-lb force in the y and z directions, respectively. The gear also imparts a 1000 in-lb torque on thew shaft.

(a) Determine the left and right reactions.

(b) Find the bearing load required to achieve a 5 million revolutions life.

It is best to break down the shaft diagram into a FBD for the y and z directions. Figures 4.19 and 4.20 are the FBD for the y and z directions, respectively. The torque diagram is illustrated in Figure 4.21. The diagram is not required for this problem but included to show that the constant torque assumption is not violated.

(a) Let's start by using Figure 4.19 and sum the forces in the y-direction. Next, sum the moments about the right bearing reaction:

$$\uparrow + \sum F_y = R_{Ly} + R_{Ry} - 1100 - 700 = 0$$

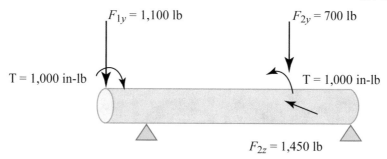

Figure 4.18: Shaft motivation problem.

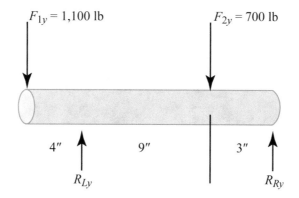

Figure 4.19: FBD y-direction.

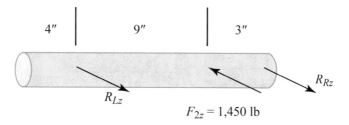

Figure 4.20: FBD z-direction.

$$\circlearrowleft \; + \sum M_{Rz} = 0$$

$$(1100)(16) - R_{Ly}(12) + (700)(3) = 0$$

$$R_{Ly} = \frac{(1100)(16) + (700)(3)}{12} = 1642 \text{ lb}$$

$$R_{Ry} = 1800 - R_{Ly} = 1800 - 1642 = 158 \text{ lb}.$$

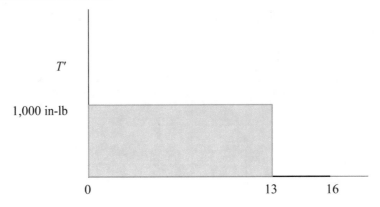

Figure 4.21: Torque diagram.

Repeating for Figure 4.20 in the z-direction:

$$\nearrow + \sum F_z = R_{Lz} + R_{Rz} - 1450 = 0$$

$$\circlearrowleft + \sum M_{Ry} = 0$$

$$R_{Lz}(12) - (1450)(3) = 0$$

$$R_{Lz} = \frac{(1450)(3)}{12} = 362.5\,\text{lb}$$

$$R_{Rz} = 1450 - R_{Lz} = 1450 - 362.5 = 1087.5\,\text{lb}.$$

We have found the reaction loads in the y and z directions. All that is left is to sum them up vectorially:

$$R_L = \sqrt{R_{Ly}^2 + R_{Lz}^2} = \sqrt{(1642)^2 + (362.5)^2} = 1682\,\text{lb}$$

$$R_R = \sqrt{R_{Ry}^2 + R_{Rz}^2} = \sqrt{(158)^2 + (1087.5)^2} = 1090\,\text{lb}.$$

(b) Use the max reaction load, $R_L = 1682$ to find C:

$$\left(\frac{L_1}{1 \times 10^6}\right) = \left(\frac{C}{P_1}\right)^k$$

$$C = P_1 \left(\frac{L_1}{1 \times 10^6}\right)^{\frac{1}{k}}$$

$$C = (1682) \left(\frac{5 \times 10^6}{1 \times 10^6}\right)^{\frac{1}{3}}$$

$$C = 2876\,\text{lb}.$$

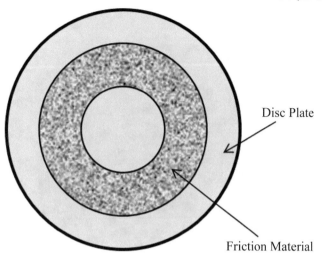

Disc Plate

Friction Material

Figure 4.22: Disc clutch.

4.10   CLUTCHES AND SPRINGS (A RELATIONSHIP)

Disc clutches and springs are typically treated as separate topics. However, in this section we will establish the relationship between the actuating force, F_a and torque. We introduce how to design/choose a spring to achieve the force as well as the stresses induced in the spring by the force. Finally, we discuss a cone clutch and its advantages and limitations vs. disc clutch. To start, Figure 4.22 illustrates the key components of a disc clutch. Figure 4.23 establishes the relationship between the actuating force (imparted by a spring and normal to the face) and the torque (imparted by the friction force).

Case I: Uniform Pressure. A constant pressure, p, acts normal to the disc clutch on the infinitesimal area, dA, shown in Figure 4.24. We can establish the mathematical relationship using Equations (4.54) and (4.55):

$$dA = rdrd\theta$$
$$dA = 2\pi rdr \qquad (4.54)$$

$$dF_n = dF_a = pdA$$
$$dF_n = dF_a = 2\pi prdr. \qquad (4.55)$$

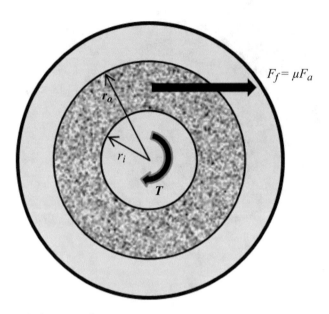

Figure 4.23: Disc clutch forces and torque.

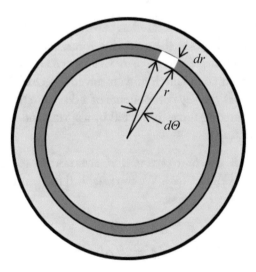

Figure 4.24: Infinitesimal area and force.

Integrating the normal (or actuation) force over the friction surface area yields:

$$F_n = \int_{r_i}^{r_o} dF_n = 2\pi p \int_{r_i}^{r_o} r\,dr$$

$$F_n = 2\pi p \left(\frac{r^2}{2}\right)\bigg|_{r_i}^{r_o}$$

$$F_n = 2\pi p \left(\frac{r_o^2 - r_i^2}{2}\right)$$

$$F_n = F_a = \pi p \left(r_o^2 - r_i^2\right). \tag{4.56}$$

To develop the relationship between the torque transmitted by the clutch and the actuating force, we recall from physics that the friction force is $F_f = \mu F_a$. An infinitesimal force, dF, acts on dA. Therefore, an infinitesimal torque, dT, is defined as $dT = r\,dF_f$:

$$dF_f = \mu\,dF_a$$

$$dT = r\,dF_f = r\mu\,dF_a$$

$$dT = r\mu(2\pi p r\,dr) = 2\mu\pi p r^2\,dr$$

$$T = \int_{r_i}^{r_o} dT = 2\mu\pi p \int_{r_i}^{r_o} r^2\,dr$$

$$T = 2\mu\pi p \left(\frac{r^3}{3}\right)\bigg|_{r_i}^{r_o}$$

$$T = 2\mu\pi p \left(\frac{r_o^3 - r_i^3}{3}\right). \tag{4.57}$$

Solving Equation (4.56) for p and substituting into Equation (4.57) yields the relationship between the actuating force and torque:

$$F_a = \pi p \left(r_o^2 - r_i^2\right)$$

$$p = \frac{F_a}{\pi \left(r_o^2 - r_i^2\right)}$$

$$T = 2\mu\pi \left(\frac{F_a}{\pi \left(r_o^2 - r_i^2\right)}\right)\left(\frac{r_o^3 - r_i^3}{3}\right) \tag{4.58}$$

$$T = \frac{2}{3}\mu F_a \left(\frac{r_o^3 - r_i^3}{r_o^2 - r_i^2}\right).$$

Case II: Uniform Wear. Typically, we are concerned with uniform wear not pressure. In this case the pressure, p, is not uniform over the friction material and p_{max} is highest at r_i. Equations (4.59) and (4.60) are the formulas for F_a and T for a disc clutch under for uniform wear.

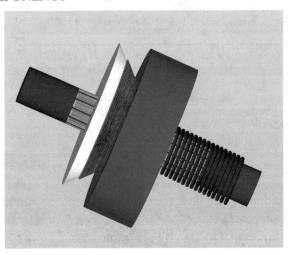

Figure 4.25: Cone clutch.

For design purposes, under uniform wear, the optimal relationship between the r_1, and r_o can be derived using Equation (4.61):

$$F_a = 2\pi p_{max} r_i \left(r_o^2 - r_i^2 \right) \tag{4.59}$$

$$T = \mu F_a \left(\frac{r_o + r_i}{2} \right) \tag{4.60}$$

$$
\begin{aligned}
F_a &= 2\pi p_{max} r_i \left(r_o^2 - r_i^2 \right) \\
\frac{dF_a}{dr_i} &= 0 \\
\frac{dF_a}{dr_i} &= 2\pi p_{max} (r_i r_o^2 - r_i^3) = 0 \\
3 r_i^2 &= r_o^2 \\
r_i &= \sqrt{\frac{1}{3}} r_o.
\end{aligned} \tag{4.61}
$$

Figure 4.25 illustrates a cone clutch. Similar to the disc clutch the equations for actuating force and torques are shown in Equations (4.62)–(4.64). It is easy to see that when setting the angle, $\alpha = 90°$ that the equations are identical to the disc clutch derived earlier. In reality the disc clutch is a special case of the cone clutch. For design purposes the angle α should be between $8 < \alpha < 12°$.

Uniform pressure:

$$F_n = F_a = \frac{\pi p \left(r_o^2 - r_i^2\right)}{\sin(\alpha)} \tag{4.62}$$

$$T = \frac{2\mu F_a}{3\sin(\alpha)} \left(\frac{r_o^3 - r_i^3}{r_o^2 - r_i^2}\right). \tag{4.63}$$

Uniform wear:

$$T = \frac{\mu F_a}{\sin(\alpha)} \left(\frac{r_o + r_i}{2}\right). \tag{4.64}$$

Example 4.9   (Disc Clutch)
Find the actuating force for a disc clutch assuming uniform wear. The clutch must transmit the torque induced from a 15 HP motor running at 1200 rpm. The outer friction surface has a radius, $r_o = 2.5$ in. and a coefficient of friction, $\mu = 0.2$:

$$HP = \frac{Tn}{63000}$$

$$T = \frac{(15)(63000)}{1200} = 787.5 \text{ in-lb}$$

$$r_i = \sqrt{\frac{1}{3}} r_o = r_i = \sqrt{\frac{1}{3}}(2.5) = 1.44 \text{ in.}$$

$$T = \mu F_a \left(\frac{r_o + r_i}{2}\right)$$

$$F_a = \frac{2T}{\mu(r_o + r_i)} = \frac{(2)(787.5)}{(0.2)(2.5 + 1.44)} = 1998.7 \text{ lb.}$$

Example 4.10   (Cone Clutch)
Compare how much torque can be transmitted by a cone clutch (assume uniform wear) given

Figure 4.26: Spring.

an angle $\alpha = 10°$ and $\alpha = 20°$. The cone clutch has:

$$F_a = 75\,\text{lb}$$
$$\mu = 0.35$$
$$r_i = 2''$$
$$r_o = 4''$$
$$T = \frac{\mu F_a}{\sin(\alpha)}\left(\frac{r_o + r_i}{2}\right)$$
$$T_{10} = \frac{(0.35)(75)}{\sin(10)}\left(\frac{4+2}{2}\right) = 453.5\,\text{in-lb}$$
$$T_{20} = \frac{(0.35)(75)}{\sin(20)}\left(\frac{4+2}{2}\right) = 230.2\,\text{in-lb}.$$

Compression Springs

We limit this section to the design and analysis of compression springs and their relationship with to clutches. Figure 4.26 illustrates a standard compression spring and Figure 4.27 shows the diameter definitions. Cutting the spring (Figure 4.28) results in the FBD illustrated in Figure 4.29. After a force is applied to the spring the wire undergoes two stress modes. The spring sees a pure shear stress defined by Equation (4.65) and a torsional stress (see Equation (2.8)). Under linear assumptions the stress can be added and the shear stress is given in Equation (4.66):

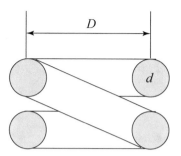

Figure 4.27: Spring diameter definitions.

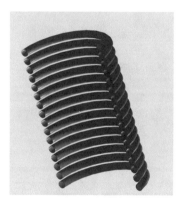

Figure 4.28: Spring cross-section.

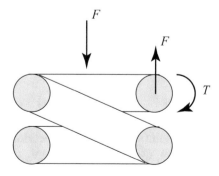

Figure 4.29: Spring FBD.

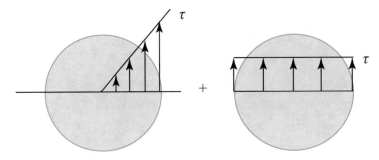

Figure 4.30: Spring-internal shear stress components.

$$\tau = \frac{F}{A} = \frac{4F}{\pi d^2} \tag{4.65}$$

$$\tau = \tau_F + \tau_T$$

$$\tau = \frac{F}{A} + \frac{Tr}{J}$$

$$J = \frac{\pi d^4}{32}$$

$$r = \frac{d}{2}$$

$$T = F\left(\frac{D}{2}\right)$$

$$\tau = \frac{4F}{\pi d^2} + \frac{\left(\frac{FD}{2}\right)\left(\frac{D}{2}\right)}{\frac{\pi d^4}{32}} \tag{4.66}$$

$$\tau = \frac{4F}{\pi d^2} + \frac{8FD}{\pi d^3}$$

define C as   Spring index $C = \frac{D}{d}$

$$\tau = \frac{8FD}{\pi d^3}\left(1 + \frac{1}{2C}\right).$$

The spring equation is given in Equation (4.67) and the deflection, δ, and spring contant, k, are defined in Equations (4.68) and (4.69), respectively, where N_a is the number of active coils in the spring, G is the shear modulus:

$$F = k\delta \tag{4.67}$$

$$\delta = \frac{8FD^3 N_a}{d^4 G} \tag{4.68}$$

$$k = \frac{d^4 G}{8 D^3 N_a}. \tag{4.69}$$

Finally, we need to account for the spring undergoing either static or fatigue loading. K is the static load factor given in Equation (4.70). K_w is the Wahl factor, used for fatigue loading, and defined in Equation (4.71). In general, the shear stress acting on a compression spring is given in Equation (4.72):

$$K = \left(1 + \frac{1}{2C}\right) \tag{4.70}$$

$$K_w = \frac{4C - 1}{4C - 4} + \frac{0.615}{C} \tag{4.71}$$

$$\tau = \frac{8FD}{\pi d^3} K \quad \text{or} \quad \tau = \frac{8FD}{\pi d^3} K_w. \tag{4.72}$$

Example 4.11 (Compression Spring)
(a) Find the stress in a compression spring under static load conditions given:

$$F = 145 \, \text{lb}$$
$$D = 1.55''$$
$$d = 0.192''$$
$$G = 11.5e^6 \, \text{psi}$$
$$N_a = 8$$

$$\tau = \frac{8FD}{\pi d^3} = \frac{(8)(145)(1.55)}{\pi (0.192)^3} = 80,860 \, \text{psi}$$

$$C = \frac{D}{d} = \frac{1.55}{0.192} = 8.07$$

$$K = \left(1 + \frac{1}{2C}\right) = \left(1 + \frac{1}{(2)(8.07)}\right) = 1.062$$

$$\tau = (80860)(1.062) = 85875 \, \text{psi}.$$

(b) Repeat the previous example using Whal factor.

$$K_w = \frac{4C - 1}{4C - 4} + \frac{0.615}{C} = \frac{(4)(8.07) - 1}{(4)(8.07) - 4} + \frac{0.615}{8.07} = 1.182$$

$$\tau = (80860)(1.182) = 95880 \, \text{psi}.$$

Table 4.6: Square keyway size, b

Shaft Diameter (inch)	Key Size, b (inch)
1/2 – 9/16	1/8
5/8 – 7/8	3/16
15/16 – 1 1/4	1/4
1 5/16 – 1 3/8	5/16
1 7/16 – 1 3/4	3/8
1 13/16 – 2 1/4	1/2
2 5/16 – 2 3/4	5/8
2 15/16 – 3 1/4	3/4
3 5/16 – 3 3/4	7/8
3 15/15 – 4 1/2	1
4 9/16 – 5 1/2	1 1/4
5 9/16 – 6	1 1/2

Example 4.12   (Disc Clutch-Spring Relationship)
In Example 4.9 we found the actuating force required to transmit a torque from a 15 HP motor running at 1200 rpm. Given the maximum deflection of the spring is 2″, find the spring constant required:

$$F_a = 1998.7\,\text{lb}$$
$$F = k\delta$$
$$k = \frac{F_a}{\delta} = \frac{1998.7}{2} = 999.4\,\text{lb/in.}$$

4.11   KEYWAYS

The last section in this chapter deals with square keyways. A keyway connects the gear, pulley, or sprocket to the shaft and plays a critical part in the system. Its failure modes must be analyzed and this section looks at failure due to shear and compression. Figure 4.31 illustrates a shaft with a keyway and Figure 4.32 shows the FBD of a keyway. Table 4.6 gives keyway size, b, as a function of shaft diameter to be used to calculate length, L. Equation (4.73) is the shear stress in a keyway and Equation (4.74) calculates the compression stress:

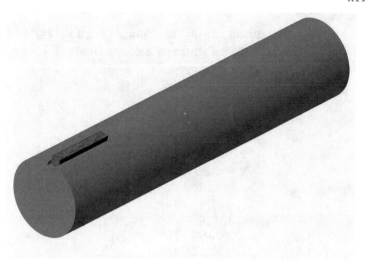

Figure 4.31: **Keyway.**

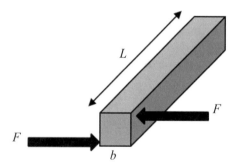

Figure 4.32: **Keyway.**

$$\tau = \frac{F}{bL} \tag{4.73}$$

$$\sigma = \frac{F}{0.5bL} \tag{4.74}$$

Remembering $T = F(\frac{D}{2})$ we can rewrite Equations (4.73) and (4.75) as a function of torque:

$$\tau = \frac{2T}{bLD} \tag{4.75}$$

$$\sigma = \frac{4T}{bLD}. \tag{4.76}$$

Example 4.13   (Keyway)

A 1/2-in. diameter shaft must transmit the torque produced from a 5 HP @ 1750 rpm motor. Using a key made from 1020 cold drawn steel and a safety factor of 2.5 determine the key's length.

Material data:

$$\sigma_u = 61{,}000 \, \text{psi}$$
$$\sigma_y = 51{,}000 \, \text{psi}$$
$$\tau = 0.5\sigma_u = 30{,}500 \, \text{psi}$$

$$HP = \frac{Tn}{63{,}000}$$
$$T = \frac{(5)(63{,}000)}{1750} = 180 \, \text{in-lb}$$
$$b = 1/8 \quad \text{from Table 4.6.}$$

Shear:

$$\tau = \frac{2T}{bLD}$$
$$L = \frac{2T}{b\tau D} = \frac{(2)(180)}{(0.125)(30{,}500)(0.5)} = 0.189 \, \text{in.}$$

Using F.S. yields L as: $L = (2.5)(0.189) = 0.472 \, \text{in.}$

Compression:

$$\sigma = \frac{4T}{bLD}$$
$$L = \frac{4T}{b\sigma D} = \frac{(4)(180)}{(0.125)(51{,}000)(0.5)} = 0.226 \, \text{in.}$$

Using F.S. yields L as: $L = (2.5)(0.226) = 0.565 \, \text{in.}$

Use $L = 0.565$ in. key.

CHAPTER 5

Computational Machine Design

5.1   INTRODUCTION

This chapter provides examples of computational techniques to solve a few of the topics introduced thus far. MATLAB® and EXCEL® are powerful programs that allow tedious calculations to be automated. The goal of this chapter is not to develop commercial grade software to solve engineering problems. The goal is to provide guidance and code examples to solve engineering problems and gain an in-depth appreciation of the effort required to produce commercially available software. More important is for designers and engineers to come away with an understanding that you should only trust the answers of your code after you performed a few hand calculations. Code examples are provided for:

1. Cross-Product

2. 1D stress analysis. Axial, shear, and torsion (using Excel VBA)

3. Pulley and Sprocket reaction loads on bearing (with shear and bending moment diagrams)

4. Mohr's Circle

5. Shaft diameter sizing and the Soderberg method

6. Cone Clutch

5.2   CROSS-PRODUCT

Equation (2.4) is coded, in MATLAB, and titled: *Cross Product Torque and Bending Moment Analysis Computational Machine Design*. This is a simple code to write and run. But, the output is critical in machine design. Each component has a specific failure mode associated with it. For example, Mx is the torque applied about the x-axis and a shaft will see a shear stress compared with a force about the y and z axes which impart a bending stress:

$$\vec{M} = (r_y F_z - r_z F_y)\hat{\imath} - (r_x F_z - r_z F_x)\hat{\jmath} + (r_x F_y - r_z F_x)\hat{k}. \qquad (2.4)$$

Cross-product torque and bending moment analysis computational machine design

```
% Moment is returned representing:

% Torque=Mx;
% BendingyAXIS=My;
% BendingzAXIS=Mz;

rx=input('rx = ');
ry=input('ry = ');
rz=input('rz = ');
Fx=input('Fx = ');
Fy=input('Fy = ');
Fz=input('Fz = ');
Calculate M = r x F
Mx= (ry*Fz)-(rz*Fy);
My= -((rx*Fz)-(rz*Fx));
Mz= (rx*Fy)-(ry*Fx);

%Print to screen
Torque=Mx
BendingyAXIS=My
BendingzAXIS=Mz
```

5.3   1D STRESS ANALYSIS

Figure 5.1 is the Excel VBA Graphical User Interface (GUI) to solve 1D stress analysis problems in axial, shear, and torsion. The program solves Equations (2.7) and (2.8). Also, Equation (5.1) solves for the deflection of an axial loaded bar with a uniform cross-section and Equation (5.2) is the shear pin equation. Figure 5.2 illustrates the VBA environment required to develop forms, command buttons, etc.:

$$\sigma = \frac{P}{A}$$

$$\tau_{max} = \frac{TR}{J}$$

$$\delta = \frac{PL}{AE} \tag{5.1}$$

$$\tau = \frac{F}{A}. \tag{5.2}$$

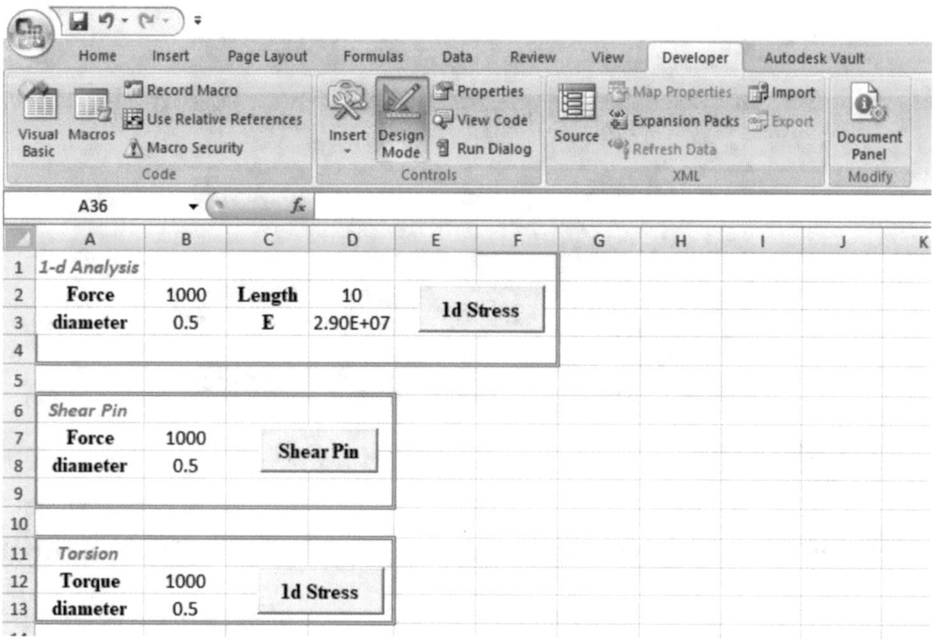

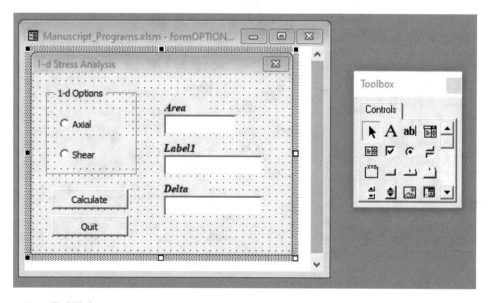

Figure 5.1: Excel VBA program for 1D stress analysis.

Figure 5.2: 1D VBA environment.

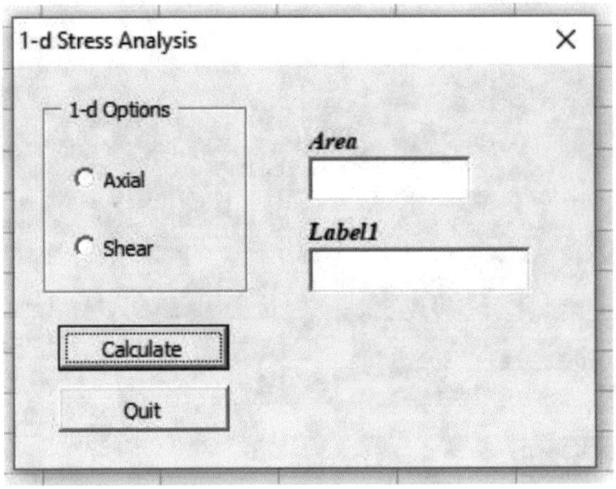

Figure 5.3: 1D options GUI.

Code attached to "Calculate" command button in Figure 5.3 to solve 1D stress analysis

```
Private Sub cmdCalculate_ Click()
force = Worksheets(3).Range("B2")
d = Worksheets(3).Range("B3")
area = 3.14159 * (d ^ 2 ) / 4
sigma = force / area
If optAXIAL.Value = True Then
Length = Worksheets(3).Range("D2")
E = Worksheets(3).Range("D3")
Delta = (force * Length) / (area * E)
lblTYPE.Caption = "sigma"
lblDELTA.Visible = True
txtDELTA.Visible = True
txtDELTA = Format (Delta, "# 0.###")
Else
lblTYPE.Caption = "tau"
lblDELTA.Visible = False
txtDELTA.Visible = False
End If
txtAREA = Format(area, "#0.###")
txtRESULTS = Format(sigma, "#0.###")
End Sub
```

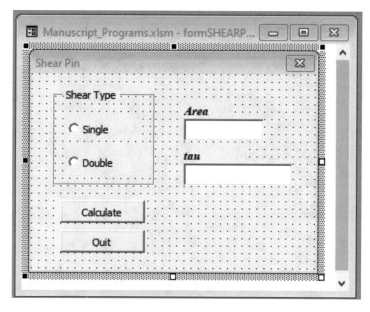

Figure 5.4: Shear pin GUI.

```
Private Sub cmdQUIT_Click()
End
End Sub
```

Code attached to "Calculate" command button in Figure 5.4 to solve shear pin

```
Private Sub cmdCalculate_Click()
force = Worksheets(3).Range("B7")
d = Worksheets(3).Range("B8")
area = 3.14159 * (d ^ 2) / 4
tau = force / area
If optSINGLE.Value = True Then
tau = force / area
Else
tau = force / (2 * area)
End If
txtAREA = Format(area, "#0.###")
txtRESULTS = Format(tau, "#0.###")
End Sub
```

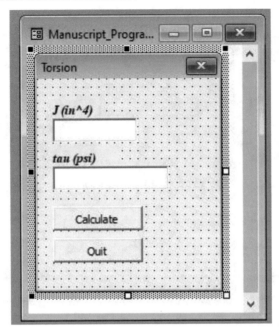

Figure 5.5: Torsion GUI.

```
Private Sub cmdQUIT_Click()
End
```

Code attached to "Calculate" command button in Figure 5.5 to solve torsion

```
Private Sub cmdCalculate_Click()
cmdCalculate.SetFocus
Torque = Worksheets(3).Range("B12")
d = Worksheets(3).Range("B13")
J = 3.14159 * (d ^ 4) / 32
tau = (Torque * (d / 2)) / J
txtJ = Format(J, "#0.#####")
txtRESULTS = Format(tau, "#0.#####")
End Sub
Private Sub cmdQUIT_Click()
End
End Sub
```

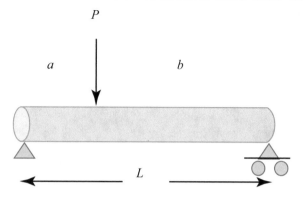

Figure 5.6: Simply supported beam.

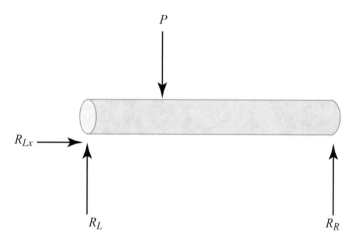

Figure 5.7: FBD for simply supported beam.

5.4   PULLEY AND SPROCKET REACTION LOADS

Figure 5.6 is a shaft supported by two bearings and subjected to simple load, P. Figure 5.7 draws the FBD for the shaft. The code solves for the right, R_R, and the left, R_L, reactions and draws the shear and bending moment diagrams, respectively:

$$R_R = \frac{Pa}{L}$$

$$R_L = \frac{Pb}{L}.$$

Code calculate reaction loads and draw shear and moment diagrams

```
% Pully/Sprocket Shaft Design Analysis

% Bearing reactions located at ends
% Input Left and Right applied load, distance along shaft,
  %and shaft length
% Returns Reaction Loads and Shear and Bending Moment Diagram
x1=input('first load located at x= ')
Ply=input('first load= ')
x2=input('second load located at x= ')
Pry=input('second load= ')
Length=input('shaft length= ')'

% Calculate the reaction loads on bearings
%
Rright=(-(Ply*x1)-(Pry*x2))/Length
Rleft=-Ply-Pry-Rright
% Plot Shear Diagram
xshear=[0,0,x1,x1,x2,x2,Length,Length];
yshear=[0,Rleft,Rleft,(Rleft+Ply),(Rleft+Ply),(Rleft+Ply+Pry),
        (Rleft+Ply+Pry),0];

%Plot Vertial Line at Origin
xstart=[-1,(Length+ 1)];
ystart=[0,0];
figure;
plot(xstart,ystart,xshear,yshear);

% Create Data for Moment Diagram
delta1=(0:(x1/10):x1);
moment1=Rleft*delta1;
delta2=(x1:(x2-x1)/10:x2);
moment2=(Rleft*delta2) + (Ply*(delta2-x1));
figure;
delta3=(x2:(Length-x2)/10:Length);
moment3=(Rleft*delta3) + (Ply*(delta3-x1))
          +(Pry*(delta3-x2));
```

```
% Plot Vertical Line at Origin
plot(xstart,ystart,delta1,moment1,delta2,moment2,
          delta3,moment3);
```

Example 5.1   (Simply Supported Beam)
Given the following loads and locations on a simply supported shaft use the program to find the reaction loads.

```
x1 = 5
Ply = -1000
x2 = 10
Pry = -1000
Length = 15
```

Output:

```
Rright = 1000
Rleft = 1000
```

Shear and bending moment diagrams (Figures 5.8 and 5.9).

5.5   MOHR'S CIRCLE

Mohr's circle is a challenging concept for students to master. The code solves for the principal stresses, σ_1 and σ_2, and $\tau_{\max}$ (Equations (3.9)–(3.11)). It is highly recommended students hand draw and calculate principal stresses to see the power of this technique and understand how to properly construct infinitesimal elements:

$$\sigma_1 = \frac{\sigma_x}{2} + \tau_{\max} \tag{3.9}$$

$$\sigma_2 = \frac{\sigma_x}{2} - \tau_{\max} \tag{3.10}$$

$$\tau_{max} = \sqrt{\left(\frac{\sigma_x - \sigma_y}{2}\right)^2 + \tau^2} \tag{3.11}$$

Code to solve Mohr's circle

```
% Mohr's Circle Code and Plot
AngleTHETA=0:360;
```

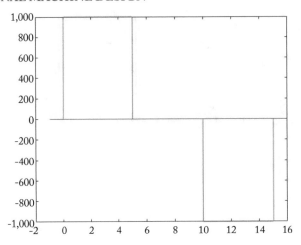

Figure 5.8: Shear plot.

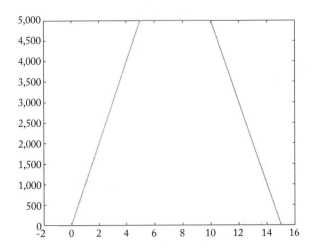

Figure 5.9: Moment diagram.

```
sigmaX=input('sigmaX = ');
sigmaY=input('sigmaY = ');
tau=input('tau = ');

% Create the Cos and Sin Matrix
MatrixCOS=cosd(AngleTHETA);
```

```
MatrixSIN=sind(AngleTHETA);
cosSQUARE=MatrixCOS.*MatrixCOS;
sinSQUARE=MatrixSIN.*MatrixSIN;
cossinMatrix=MatrixCOS.*MatrixSIN;

% Calculate the normal and tangential stress
% of the element
sigmaTHETA= sigmaX*cosSQUARE + sigmaY*sinSQUARE
            -2*tau*cossinMatrix;
tauTHETA= sigmaX*cossinMatrix - sigmaY*cossinMatrix
            + tau*cosSQUARE -
          tau*sinSQUARE;

% Find Principal Stresses
sigma1=max(sigmaTHETA)
sigma2=min(sigmaTHETA)
shearMAX=max(tauTHETA)

% Calculate the vonMises stress
sigmaV=sqrt(sigma1.^ 2 -(sigma1*sigma2)+sigma2.^ 2)
sigmaV=sqrt(sigma1*sigma1 -(sigma1*sigma2)+sigma2*sigma2)

% plot sigma vs. tau
scatter(sigmaTHETA,tauTHETA);
hold on
xline=[sigmaX,sigmaY];
yline=[tau,-tau];
line(xline,yline,'Color','red','LineStyle','--');
hold off
```

Example 5.2   (Mohr's Circle)
Given the following element loads use the code provided to solve for principal stresses:

```
sigmaX = 100
sigmaY = 50
tau = 50
```

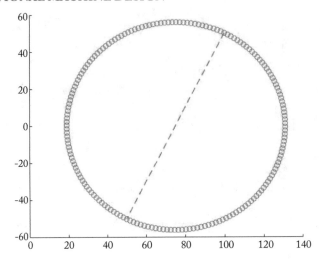

Figure 5.10: Mohr's circle.

Output:

```
sigma1 = 130.8990
sigma2 = 19.1010
shearMAX = 55.8990
sigmaV = 122.4708
```

Mohr's circle (drawn in Figure 5.10).

5.6 SHAFT SIZING AND SODERBERG METHOD FOR FATIGUE

Equations (3.22) (calculates shaft diameter) and (3.23) (endurance stress) are modeled using the code provided:

$$d \geq \left(\frac{32N}{\pi \sigma_y} \sqrt{\left(\frac{\sigma_y}{\sigma_e} M\right)^2 + \frac{3}{4}T^2} \right)^{\frac{1}{3}} \tag{3.22}$$

$$\sigma_e = \left(\frac{1}{2}\sigma_u\right) k_a k_b k_c k_d k_e k_f \frac{1}{K_f}. \tag{3.23}$$

Code to solve for shaft diameter and plot Soderberg curve

```
% Fatigue Analysis Modeling

kFACTORS= input('input all 7 k values in bracket [ ] form')
kFACTORS;

% Check of input values
ka= kFACTORS(1)
if ka > 1
warning("ka Value > 1")
end
kb= kFACTORS(2)
if kb > 1
warning("kb Value > 1")
end
kc= kFACTORS(3)
if kc > 1
warning("kc Value > 1")
end
kd= kFACTORS(4)
if kd > 1
warning("kd Value > 1")
end
ke= kFACTORS(5)
if ke > 1
warning("ke Value > 1")
end
kf= kFACTORS(6)
if kf > 1
warning("kf Value > 1")
end
kF=kFACTORS(7)
if kF < 1
warning("kF Value < 1")
end

% Input Yield and Ultimate stress
sigmaU=input('Ultimate stress= ')
```

```
sigmaY=input('Yield stress= ')
FS=input('Factor of safety=')
M=input('Bending Moment= ')
T=input('Constant Torque ')

% Basic Program where T and M are known
%
% Check sigmaU $>$ 100,000
% If greater then 100000 then set to 100000 psi for
% endurance stress calculation
if sigmaU > 100000
sigmaU= 100000;
end

% Calculate endurance stress
sigmaE=(0.5*sigmaU*ka*kb*kc*kd*ke*kf)/kF

% Calculate Shaft Diameter
coeff1=(32.0*FS)/(3.14159*sigmaY);
coeff2=((sigmaY*M)/sigmaE).wedge2;
coeff3=(0.75*T. .wedge2);
diameter3=coeff1*sqrt(coeff2+coeff3);
diameter=diameter3.wedge(1/3)
%Plot of Soderberg Fatigue Curve
%Plot Shear Diagram
xsoderberg=[0,sigmaY];
ysoderberg=[sigmaE,0];
plot(xsoderberg,ysoderberg);
title("Soderberg Fatigue Curve")
xlabel('Mean Stress')
ylabel('Alternating Stress')
```

Example 5.3   (Shaft Diameter)
Given the following data calculate shaft diameter and plot fatigue curve:

```
ka = 0.9000
kb = 0.9000
kc = 0.9000
```

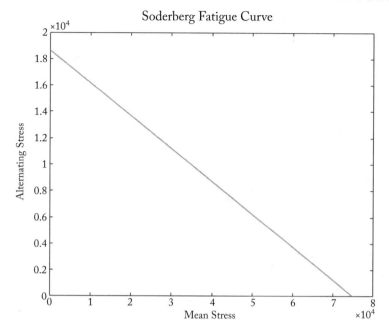

Figure 5.11: Soderberg fatigue curve.

```
kd = 1
ke = 0.8000
kf = 0.8000
kF = 1.2500
sigmaU = 125000
sigmaY = 75000
FS = 2
M = 28000
T = 5000
```

Output:

```
sigmaE = 1.8662e+04
diameter = 3.1274
```

Soderberg fatigue curve (drawn in Figure 5.11).

5.7   CONE CLUTCH

The last section developed code to solve for the actuating force, F_a, required when analyzing a cone clutch assuming uniform wear (Equation (4.64)):

$$T = \frac{\mu F_a}{\sin(\alpha)} \left(\frac{r_o + r_i}{2} \right) \qquad (4.64)$$

Code to solve for cone clutch

```
% Cone Clutch Analysis

HP=input('HorsePOWER= ');
n=input('Speed (rpm)= ');
coeffmu=input ('coefficient of friction= ');
angleALPHA=1:20;
radiusi=input('inside radius <enter 0 if unknown> = ');
radiuso=input('outside radius <enter 0 if unknown> =');

% Calculate Torque
if radiusi==0
radiusi=(sqrt(1/3))*radius
else
radiuso=radiusio/sqrt(1/3)
end
T=HP*63000/n
Fa=(2*T)*sind(angleALPHA)/(coeffmu*(radiusi+radiuso))

% Check Disc Clutch
FaDISC=(2*T)/(coeffmu*(radiusi+radiuso))
```

CHAPTER 6

Capstone SE Project

6.1   THE CAPSTONE APPROACH

The final chapter demonstrates how SE models and machine component analyses learned are incorporated into a capstone project. First, teams were formed and they were assigned to develop a conveyor system to a set of requirements. Table 6.1 are the eight design requirements at various stages of the design's life-cycle. Next, teams used brainstorming techniques to develop several design concepts. After the team choose a concept, they applied the FUSE-FFBD Model (Figure 6.1) to identify all success modes.

The goal of the capstone was for teams to identify how to meet the given requirements and not add unnecessary components or features to the design. The requirements given, in some cases, were ambiguous (Requirement 6: No custom parts) to challenge the teams to ask "what does that mean" or to let the team make a suggestion that slight modifications to an "off-the-shelf" component should be considered. Designs were required use a clutch but not beholden to a particular type. The conveyor belt speed and motor speed required a gear train (or speed reduction design). The operation and maintenance requirement tied into properly sizing bearings. Team could argue a smaller life bearing and replacement could be considered given a 2-week downtime (see requirement 7 stating 50 weeks/year).

To start, teams developed a hierarchy structure (see Figure 1.10) and identified three major subSYSTEMS. Figure 6.2 illustrates the hierarchy structure for the Power subsystem. The Power

Table 6.1: Design requirements

Requirements	Phase	Comments
Motor	Design	1.5 HP @ 860 rpm
Clutch	Design	Clutch required
Transmission	Design	30″ × 30″ footprint
Operation	Oper and Maint	6′ distance to conveyor
Conveyor	Design	4″ standard drum and 16″ conveyor belt
Manufacturer	MFG	No custom parts
Operation	Oper and Maint	50 weeks, 6 hr/day, 5 days/week
Operation	Oper and Maint	Conveyor speed: 150–200 ft/min

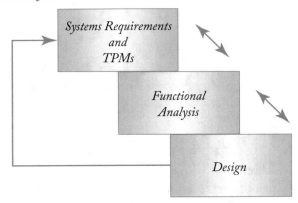

Figure 6.1: **FUSE-FFBD** model.

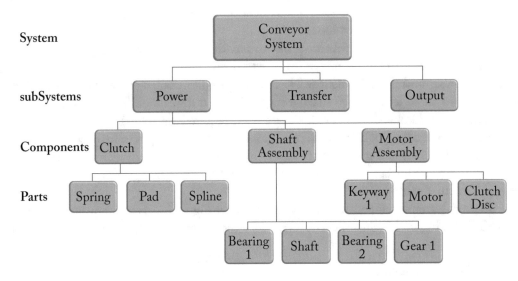

Figure 6.2: Design hierarchy structure.

subsystem contains the clutch, shaft assembly, and motor assembly. Each assembly contains its part. For example, the clutch contains the spring, disc pad, and spline. Teams used the FFBD (Figure 6.3) to identify all of the success modes. Table 6.2 identifies all of the success modes for a conveyor system concept.

Finally, teams were able to perform design calculations on all components using the techniques learned. Table 6.3 provides a sample of the shaft size and bearing life for the design illustrated in Figure 6.4.

Functional Flow Block Diagram

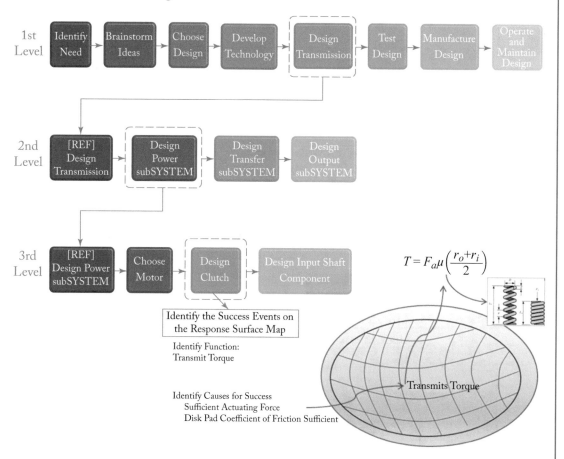

Figure 6.3: **Clutch design.**

Table 6.2: Success modes

subSystem	Part	Success Mode	Value
Power	Motor	N/A off-the-shelf	1.5 HP 860 rpm
	Clutch, disc	N/A off-the-shelf	N/A
	Clutch, pad	Sufficient friction	Design
	Clutch, spring	Sufficient spring constant	8<C<12
	Spline	Sufficient torque transfer	Design
	Shaft	Sufficient diameter	Design
	Bearing 1	Sufficient life	Design
	Bearing 2	Sufficient life	Design
	Gear 1	Sufficient strength	Design
	Keyway 1	Sufficient length	Design
Transfer	Keyway 2	Sufficient length	Design
	Gear 2	Sufficient strength	Design
	Keyway 3	Sufficient length	Design
	Gear 3	Sufficient strength	Design
	Bearing 3	Sufficient life	Design
	Bearing 4	Sufficient life	Design
	Shaft, interm	Sufficient diameter	Design
Output	Keyway 4	Sufficient length	Design
	Gear 4	Sufficient strength	Design
	Bearing 5	Sufficient life	Design
	Bearing 6	Sufficient life	Design
	Shaft, output	Sufficient diameter	Design
	Pulley 1	N/A off-the-shelf	N/A
	Flat belt	Sufficient strength	Design
	Pulley 2	N/A off-the-shelf	N/A

Table 6.3: Partial results

Requirements	Part	Design Value
Motor	Motor	110 in-lb
Clutch	Clutch, disc	N/A
Clutch	Clutch, pad	0.25
Clutch	Clutch, spring	370 lb
Trans	Shaft	0.75″
Trans	Bearing 1	768.2 10^6 rev
Trans	Bearing 2	768.2 10^6 rev

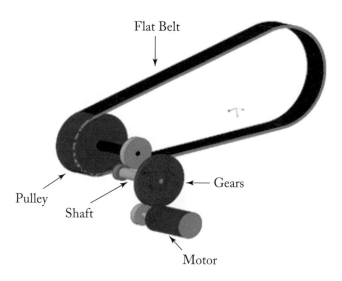

Figure 6.4: Typical design.

6.2   CONCLUSION

The capstone project was intended to give students a process for designing complex systems and an understanding that requirements might be ambiguous. The SE process allows a structured approach to design and the teams worked well together. Students were required to write and turn-in a final report to meet the requirement of written communication. Finally, the final design review was team-to-instructor instead of teams presenting their designs as in-class presentations.

Bibliography

[1] Integration Definition for Function Modeling (IDEF0), *FIPS 183*, Department of Commerce, Gaithersburg, MD, 1993. 5

[2] Hutcheson, R., McAdams, D., Stone, R., and Tumer, I. Function-Based Systems Engineering (FUSE), *Proceedings of the International Conference on Engineering Design*, pp. 1–12, Paris, France, August 28–31, 2007. 7

Author's Biography

ANTHONY D'ANGELO, JR.

Anthony D'Angelo, Jr. is a licensed Professional Engineer in the State of New Jersey and holds a B.S. and an M.Sc. in Mechanical Engineering from the New Jersey Institute of Technology and a MBA from Rutgers University. He has over 37 years of professional experience in government and industry conducting structural, thermal, and reliability analyses and authored or co-authored nine papers in the area of reliability-based design analysis and systems engineering. The author presently is conducting research, pursuing a Ph.D. in systems engineering from Colorado State University, on developing a reliability-risk modeling-based trade study tool. In addition, he is an adjunct instructor and has taught numerous courses in machine design, statics, dynamics, mechanics of materials, mechanisms, and material science at the County College of Morris, and various math courses at several County Colleges in New Jersey.

Printed in the United States
by Baker & Taylor Publisher Services